Moral Respect, Objectification, and Health Care

Meredith Celene Schwartz

Moral Respect, Objectification, and Health Care

palgrave
macmillan

Meredith Celene Schwartz
Department of Philosophy
Ryerson University
Toronto, ON, Canada

ISBN 978-3-030-02966-1   ISBN 978-3-030-02967-8   (eBook)
https://doi.org/10.1007/978-3-030-02967-8

Cover illustration: © Melisa Hasan

This Palgrave Pivot imprint is published by the registered company Springer Nature Switzerland AG
The registered company address is: Gewerbestrasse 11, 6330 Cham, Switzerland

Acknowledgements

My academic thinking has been deeply shaped by the work of Susan Sherwin who was my supervisor and continues to offer guidance and support. I am grateful to Dr. Sherwin for reading an early draft of this book. Adam Auch, Alex Wellington, Anna Macdonald, Jo Kornegay, Jordan Wadden, Khadija Coxon, Laurenne Ava Kredentser, Madelaine Ley, Mara Marin, Mitchell Wideman, Robert Murray, Rosemary Wharton, Tim Mt. Pleasant, and Victor Bruzzone all read early drafts of some of the chapters and I benefitted greatly from their feedback. I have also presented versions of these chapters at the International Feminist Approaches to Bioethics conference and the Canadian Society for Bioethics conference and received useful feedback from the audiences. Jordan Wadden was my research assistant for part of the writing process for this book, and he offered invaluable advice on restructuring some of the arguments.

I would like to thank my parents, Mary Ann and Paul Schwartz in whose back yard most of this book was written. I benefitted greatly from our conversations and their comments on multiple drafts as I was writing. My stay at their home was a perfect demonstration of the second-person care-respect that I described in the chapters of this book.

I am also grateful to my partner, Joel Swedburg, who supports me through my worst angst-ridden moments of writing and my most joyous moments of living.

I would most especially like to thank Kelly Oliver whom I met in October 2017. I had been casting about in need of guidance and meeting Dr. Oliver helped me to set my course. I was deeply moved by how much interest Dr. Oliver took in proving me with mentorship even though we had only just met. Without Dr. Oliver's guidance on strategies for academic publishing, this book would never have materialized. Dr. Oliver is a truly generous philosopher, not only with her own students and junior faculty but for even those junior faculty members she has just met.

All of the errors, omissions, and oversights remain my responsibility alone.

Contents

Abbreviations

ACT   Assertive Community Treatment
hCG   Human Chorionic Gonadotropin
HSG   Hysterosalpingogram
NT   Nuchal Translucency

1

Introduction

Abstract Schwartz fills an important gap in existing health care ethics literature by describing an egalitarian conception of moral respect which applies to autonomous and non-autonomous patients alike. It reframes questions about respect, from its target to the role that respect plays in our moral lives. Taking into account various forms of objectification, it suggests that the unique role of moral respect is to recognize a person as more than a mere object; to recognize them as an equally intrinsically valuable being who possesses dignity. Schwartz describes various forms of objectification and considers three cases in which patients are disrespected even while the doctor is upholding their autonomous decision-making.

Keywords Moral respect · Objectification · Dignity ·
Second-person standpoint · Respect for autonomy · Informed consent

© The Author(s) 2019
M. C. Schwartz, *Moral Respect, Objectification, and Health Care*,
https://doi.org/10.1007/978-3-030-02967-8_1

1.1   Introduction

My aim is to provide an egalitarian account of moral respect as it applies in health care contexts. The concept of respect is central to medical ethics. Some discussion of 'respect' has been part of medical ethics since the early days of the discipline (e.g. *The Belmont Report* 1979). The dominant discussion of 'respect' in the context of health care ethics focuses on 'respect for autonomy' and this conception is well-entrenched in legal frameworks in a number of countries. The conception of 'respect for autonomy' leaves a gap in medical ethics because it does not apply to those patients who are not autonomous. In the seventh edition of *The Principles of Biomedical Ethics*, Beauchamp and Childress explicitly write that the principle of respect for autonomy does not include those who are not autonomous and cannot be rendered autonomous, such as children or patients diagnosed with late-stage Alzheimer's disease. They state that those who are not autonomous are still covered by the concept of moral respect (Beauchamp and Childress 2013, p. 108). The concept of moral respect remains unanalysed in their edition, however. I fill this gap by reframing questions about moral respect around the role that the concept plays in our moral lives. I suggest that the fundamental role of moral respect is to recognize a person as more than a mere object, as an entity with intrinsic value of dignity rather than merely use-value of price.

To further explore the role of respect as recognizing someone as more than a mere object, I consider what it means to treat someone as an object. I suggest that disregarding someone's autonomy or violating their autonomy is one important way of treating someone as a mere object. So respect for autonomy will remain central on my account. There are, however, a number of additional ways that we can reduce someone to a mere object and I pay particular attention to these additional forms of objectification.

I describe three aspects of the concept of respect: the grounds of respect, the target of respect and the behaviour that enacts respect. The ground or basis of respect tells us why respect is owed or warranted. The target of respect identifies what sorts of things must be respected.

The behaviour that enacts respect describes how respect is put into practice in respectful relationships (Dillon 2010). A concept of moral respect should be egalitarian in the sense that it should apply equally to all persons, or members of the moral community. I suggest that we should understand respect as a concept that is grounded in dignity, which recognizes both the absolute moral value of individuals as well as the concrete particularity of their perspectives. As a result, the target of respect on this view is the person themselves, rather than an abstract feature or fact about the person. Respect is enacted through interactive second-person asymmetrical relations that treat others as more than mere objects. We disrespect others when we reduce them to mere objects, and so I consider what it is to objectify someone. There are a number of way to treat someone as an object, so the account of respect that I offer is relational and pluralist.

1.2   Respect for Autonomy and Informed Consent

The most common and legally-entrenched conception of 'respect' in health care ethics and law focuses on respect for autonomy, where autonomy is understood as making informed, voluntary choices about particular medical treatments or other aspects of care. This kind of autonomy is protected by ensuring that patients are well-informed about their condition and treatment options and that patients can give voluntary consent or refusal to these treatments. This conception of respect and its protection through informed consent is important and deserves its central place in medical practice and health law. Nothing in my discussion would undermine the importance of respect for autonomy. Respecting the autonomous decisions of competent patients is a necessary component of respecting them. Violations or denials of autonomy remain one important way that the moral equality of autonomous patients can be denied. Instead, my claim is that respecting autonomous decision is not sufficient for moral respect in health care settings. There are two reasons that it is not sufficient.

First, it is inegalitarian because it does not apply to non-autonomous patients. Second, there are cases where even autonomous patients can arguably be disrespected even while their choices about medical treatments are accepted and they have given competent informed consent. To make these two criticism clear I begin by describing Beauchamp and Childress' influential discussion of respect for autonomy. They themselves note that this conception of respect does not apply equally to all patients. The second claim will be developed throughout this chapter and the rest of the book.

On Beauchamp and Childress's account, the grounds, object, and obligations of respect all centre on capacities for autonomous decision-making and supporting patients in making autonomous decisions. The understanding of 'autonomy' that Beauchamp and Childress invoke focuses on localized, specific instances of *autonomous choice*, rather than a broader conception that understands 'autonomy' as abilities, skills, or traits of the person (2009, p. 100; 2013, p. 102). Beauchamp and Childress select 'autonomy' as the grounds of our obligations of respect in health care contexts because they want "to be as precise as possible about what is and must be respected" (Beauchamp and Childress 2009, p. 70; 2013, p. 68). They eschew the language of *respect for dignity* and *respect for persons* because they believe that the terms 'dignity' and 'person' are vague and inherently contestable (2009, pp. 66, 69–70; 2013, pp. 65, 68).

The object of respect is also autonomous decision-making on Beauchamp and Childress's view. That is, the presence of competence and other capacities for autonomous choice both explains why we must respect autonomous decisions, and autonomous decisions are the target of our respect. Since the early editions of *Principles of Biomedical Ethics* Beauchamp and Childress have contended that when we show respect to patients, the target of our respect is their choice; however, the obligations of respect have strengthened over subsequent editions. In the first edition, Beauchamp and Childress tell us, "To respect autonomous agents is to recognize with due appreciation their own considered value judgements and outlooks even when it is believed that their judgements are mistaken. To respect them in this way is to acknowledge their right to their own views and the permissibility of their actions based on such beliefs" (1979, p. 58). These obligations focused on granting persons the right to their own views, and the negative prohibition on interfering with their liberty. In more recent editions, Beauchamp and Childress have strengthened these requirements considerably and now require positive elements that acknowledge "the value and decision-making rights of persons and [enable] them to act autonomously" (2009, p. 103; 2013, p. 107). This involves "respectful *action*, and not merely a respectful *attitude*" (Beauchamp and Childress 2013, p. 107; emphasis in original). The respectful actions include such things as providing information, building up the capacity for autonomous choice, and dealing with emotions, such as fear, or other conditions that might distort autonomous actions. In contrast, disrespect for autonomy involves actions and attitudes that "ignore, demean, or are inattentive to others' rights of autonomous action" (Beauchamp and Childress 2009, p. 103; 2013, p. 107). On this view, the behaviours and attitudes associated with both respect and disrespect focus on autonomous decision-making.

Although Beauchamp and Childress's account of 'respect' is clearer than some of the discussions of respect found within general mainstream bioethics, this clarity is bought at a price when we consider the scope of respect—that is, to whom (or what) respect is owed. Because Beauchamp and Childress take autonomy as the object of respect, on their view our obligations related to this kind of respect extend only to those who are autonomous or those who have expressed their wishes through advance directives. They write, "Our obligations to respect autonomy do not extend to persons who cannot act in a sufficiently autonomous manner (and who cannot be rendered autonomous) because they are immature, incapacitated, ignorant, coerced or exploited. Infants, irrationally suicidal individuals, and drug-dependent patients are examples" (2009, p. 105). In the seventh revision to their view, Beauchamp and Childress have added the sentence "This standpoint does not presume that these individuals are not owed moral respect. In our framework, they have a significant moral status… that obligates us to protect them from harm-causing conditions and to supply medical benefits" (2013, p. 108). Interactions with those who cannot act sufficiently autonomously will still be subject to the remaining three principles of beneficence, non-maleficence, and justice. Beauchamp and Childress offer a vague suggestion that non-autonomous patients will still deserve some form of respect. This other form of respect, that Beauchamp and Childress call *moral respect* remains unanalysed in their book, however. I infer that Beauchamp and Childress recognize that their account of respect is inegalitarian in that it treats autonomous and non-autonomous patients differently while moral respect is egalitarian and would require respect for patients regardless of their decision-making capacities. This latter form of egalitarian moral respect is my focus in this book.[1]

1.3   Cases of Disrespect?

My second criticism, that respect for autonomy is not sufficient for moral respect even among autonomous patients, is more difficult to establish. To provide a sketch of my concern, I present several cases where patients describe the experience as disrespectful but not as a violation of autonomy.

1.3.1 Case A: A Child with Down Syndrome[2]

A mother and her 5-year-old daughter with translocation Down syndrome came into the high-risk pregnancy clinic for an ultrasound. The nuchal translucency (NT) scan results were abnormal and showed an increased probability that the foetus she was carrying also had this heritable form of Down syndrome. The obstetrician explained the results and recommended follow-up testing and explained the relative risk of miscarriage compared to the risk the foetus had Down syndrome.

---

[1]Throughout the book if I use 'respect' this should be understood to mean 'moral respect' unless I specify another concept such as "respect for authority" or "respect for autonomy."

[2]Chapter 3 will return to this case to consider moral respect for non-autonomous children. Some details of this case have been changed or omitted to protect the identities of those involved.

He explained her options including the possibility of further testing, which could be performed at the hospital and booked at the front desk. As he spoke, the mother became increasingly tense and began to pull her daughter closer. She kept insisting, "My baby's fine," which prompted the obstetrician to begin another round of explanation with increasing contextualization. When the mother left, the doctor commented to me, "She is in denial. I couldn't get through to her. She was being very difficult." Later, the mother told me, "the doctor was so disrespectful to my daughter."

1.3.2 Case B: A Young Mother with Choriocarcinoma[3]

At age 22 Tara[4] was diagnosed with stage IV choriocarcinoma (placental cancer) after the birth of her first child. The cancer had caused haemorrhaging, which required a hysterectomy performed by her gynaecological oncologist. On a subsequent visit to the gynaecological oncologist several weeks later the doctor asked Tara whether she was using birth control. She said she was not, and he began to explain to her how important it was to ensure that she did not get pregnant while she was being treated for choriocarcinoma. He explained that the chemotherapy she was receiving could cause abnormalities in a foetus. Further, since choriocarcinoma is placental cancer, pregnancy and the resulting placental development could make it difficult to track whether the treatments were working. As he spoke Tara lost her usual sunny demeanour and began to grit her teeth and flare her nostrils. After the appointment, Tara said to me, "Wow! I have never felt so disrespected in my whole life" (Johnson and Schwartz 2007).

---

[3]Chapter 2 returns to *Case B* as I consider the ways in which medical epistemology and clinical care are objectifying.

[4]This is her real name, and I have permission to discuss her story in the context of this book. Tara Johnson and I wrote about her experience in *Gestational Trophoblastic Neoplasia* (2007). Some details of this case have been changed or omitted to protect the identities of those involved.

1.3.3 Case C: Fainting and Fertility Tests

In my thirties, I underwent testing for infertility. I received written information from my fertility specialist, she offered to have a more in-depth discussion if I had any questions. The information about the hysterosalpingogram (HSG) said there might be some cramping, similar to a menstrual period, or some dizziness. The sheet said that some women like to bring their partner to accompany them after the test in the rare case of fainting. I did not think this would be necessary for me, as my menstrual cramps are usually minor and so I thought I would be fine. The test was more painful than I anticipated. As I was trying to get dressed afterward I got dizzy and felt faint. I couldn't stand up to get my underwear on, so I crawled out of the changing room still in the gown and nothing else. There were some chaises lounges in the room, but there was no one to ask whether I could lie down. Out of necessity I collapsed on the couch as I felt very faint and dizzy. I felt abandoned in the empty room. I was alone until a custodian walked in and I asked him for a cup of water. The experience made me feel disrespected and I decided I would never return to that clinic.[5]

1.3.4 Case D: An Elderly Woman with Alzheimer's Disease[6]

Olive was eighty years old when she started to get lost and confused. When she found she did not know where she was she would sometimes get frightened and anxious. Within a few years her care became more than her husband, Alfred, could handle. Alfred and Olive discussed moving her to a long-term care home where she could get the support she needs and Alfred could visit her several times a week.

---

[5]Some details of this case have been changed or omitted to protect the identities of those involved. I return to this case in Chapter 4.

[6]This case is an amalgam of a fictionalized case described by Tom Kitwood based on his observations (2011, pp. 91–94) and descriptions of observations made in an Ontario residential care unit provided by Moira Welsh (2018). Chapter 3 returns to a discussion of respect and elderly people who are no longer autonomous.

They chose a good home with good measures of performance and quality of care. The staff at the home kept to a meticulous schedule of feeding, cleaning and recreation; everything was well documented and tracked. During meals the staff were often in a hurry so they fed Olive quickly while they chatted happily to each other. Each night they bathed Olive and put her to bed. Olive was often upset at night and one evening she bit the nurse who was checking on her. The staff asked Alfred whether he would approve the use of sedatives to keep Olive calm and he agreed. Olive spent her days watching television, but gradually began to find it hard to follow the stories, which increased her distress and made her agitated. One day Olive was wandering around the locked ward saying quietly, "I'm in a cage" (Welsh 2018). Another resident got angry and started shouting at Olive and she hit the resident causing bruises (Kitwood 2011, p. 93). After consulting with Alfred, the staff decided to give Olive sedatives during the day as well as at night. Olive's days consisted of a routine of being awakened, washed and fed then she was put in a chair and spend most of the day looking out the window. She occasionally got up to wander around the ward and then by 8 pm she was put to bed. After a few months, Olive stopped recognizing Alfred when he visited. His visits then became less frequent as he did not see any purpose in visiting someone who did not know him. Olive spent the end of her days looking out the window in the care home.

1.3.5 Preliminary Case Analysis

The first three of these cases involved autonomous, competent patients who were given adequate information and allowed to make a voluntary choice free of coercion. The last case involves a patient who no longer has the capacity to make treatment decisions, but has a substitute decision-maker in her loving husband. None of these cases involved denials or violations of autonomy. Nevertheless, in each of the cases the patient involved described the experience as disrespectful or demonstrated distress in the situation. In the first case the obstetrician was in the process of providing the mother with complete and accurate information about the "risk" that her foetus had Down syndrome. When she insisted that her baby was fine, the obstetrician interpreted this as a lack of understanding and began to explain the risks and options over again. He was not coercing her choice or trying to get her to choose one option rather than another. If the interaction was disrespectful, this was not the result of a violation of informed consent law or the mother's autonomy. Nevertheless, the mother told me she felt the doctor disrespected her daughter during the encounter. Tara's doctor was also in the midst of providing accurate information to Tara. The information he was providing was information that most patients would find highly relevant to their decision-making, since most patients would want to avoid medications during pregnancy that are known to cause foetal abnormalities. Tara described this interaction infantilizing, however. She said she thought the doctor was talking down to her and demeaning her. In my own case, there was no decision to be made at the moment when I felt disrespected. I had already made the free and informed choice to consent to a HSG. The procedure was already over, there was no further medical intervention that needed to be done, at least not at that moment. I had misjudged my experience of the procedure, it hurt considerably more than I was anticipating, but it was not the misjudgement that made me feel disrespected. I did not think I was misinformed about the procedure, nor did I regret my decision regarding the procedure. Olive's needs were met and she did not face abuse or neglect from the staff. Although Olive no longer had the capacity to consent to treatment regimens, Alfred provided consent on her behalf. If there is any disrespect happening in these cases—that is, if the patients involved are right to describe the experience as treating them or their child as less morally valuable, or reducing them to mere objects—the disrespect is not the result of a denial or violation of autonomy. Of course, the mere fact that a patient does not feel respected does not entail that disrespect actually occurred. We can be wrong in our interpretation of an event. In order to understand whether these interactions were disrespectful we need a more thorough discussion of what it means to say someone has "equal moral value" and what it means to treat someone as a "mere object." My aim in this book is to describe an account of moral respect that applies to both autonomous and non-autonomous patients alike and that describes a variety of ways of recognizing and treating others with dignity. I provide a brief overview of the account in the next section.

1.4   Summary of the View

Throughout this book I describe an egalitarian account of moral respect as a second-person relation that recognizes and affirms the moral dignity of others and treats them as an equally valuable member of the moral community. I argue that recognizing the equal value of persons entails treating them as more than a *mere* object. The central and unique role for moral respect is this recognition of moral equality and no other concept captures just this element. I draw upon Stephen Darwall's (2006) discussion of respect as a second-person concept that involves an engagement from, what Peter Strawson (1974) calls, the interactive stance. This differs from moral judgements that take place from third-person perspective or the objective stance.

My account differs from Darwall's because he focuses on reactive attitudes and holding each other to account from positions of equal moral authority. While I agree these are important aspects of moral behaviour, I do not believe that they exhaust moral behaviours from the second-person standpoint. In addition to offering moral reasons, reactive attitudes, and accountability we also engage with each other second-personally when we provide care-respect (Dillon 1992a, b), support each other's agency (Kontos 2005; Kitwood 1990, 1993), offer narratives to explain who we are and whom we take others to be (Baylis 2017), and when we engage with others in ways that recognize their equal moral value as more than a *mere* object. In particular, my account of moral respect differs from Darwall's account because he focuses on equal moral status and authority, whereas I argue that moral respect is grounded in moral dignity and that one need not have equal status or authority to have the equal value named by moral dignity.

Given the role of respect in recognizing someone as more than a mere object I consider a variety of forms of objectification drawing on the work of Martha Nussbaum, Rae Langton, and Kay Toombs, who offer at least ten ways in which we can objectify others. First, we treat people as objects when we use them *instrumentally* as a tool for our own ends and desires (Nussbaum 1995, p. 257). Second, when we deny or violate someone's autonomy we treat them as though they lack the capacity for self-determination (Nussbaum 1995, p. 257), or as though their autonomous capacities and expressed wishes don't need to be taken into account (Langton 2005, p. 246). Third, people have agency, which is thoroughly embodied, and we treat someone as an object when we treat them as *inert* or lacking agency (Nussbaum 1995, p. 257). When Kant describes the concept of moral dignity as a kind of moral value, he notes that things with moral dignity are non-fungible. They should not be treated as though they are interchangeable with other people. Treating others as *fungible* is a fourth way for them to be objectified (Nussbaum 1995, p. 257). Something that is *violable* lacks boundary integrity and can be broken, cut-up or destroyed (Nussbaum 1995, p. 257). When we treat people as violable we objectify them in a fifth way. A sixth way to treat someone as an object is to treat the person or parts of the person as though they could be *owned* (Nussbaum 1995, p. 257). Seventh, people have subjectivity which involves their experiences and emotions. We treat people as objects when we treat them as though they lack these experiences (Nussbaum 1995, p. 257), or as though these experiences and emotions exist but don't matter (Langton 2005, p. 246). Eighth, when we *Reduce a person to their body or its parts* we objectify the person by identifying them with their body (Langton 2005, pp. 246–247). Someone is *silenced* when they are treated as though they cannot speak, or as though their speech is irrelevant or illegitimate, which is a ninth form of objectification (Langton 2005, p. 247). Finally, a person is *alienated* from their own body when one's body comes to feel as though it were an object or something "other-than-me" which is a common aspect of experiences of serious illness (Toombs 1988, pp. 215–216).

I focus on these forms of objectification and I argue that medical epistemology, research and care, inherently involve some of these forms of objectification. In addition, phenomenological experiences of illness involve many of these forms of objectification and patients can come to find themselves alienated from their bodies. Where once their bodies were the centre of their world and agency, illness can cause a rift in their embodiment. Patients often begin to think of their body as an object, as something limiting rather than facilitating their agency. Because the experience of illness and treatment is inherently objectifying moral respect deserves its place as a central moral concept within health care.

My account draws elements from Immanuel Kant, but it is not a Kant ian account. In particular, I draw from Kant (1996) his discussion of moral dignity as a concept that names a particular kind of moral value. Kant describes moral dignity as an intrinsic, categorical, absolute value that is inalienable and applies to non-fungible beings. In contrast, price is an extrinsic, contingent, relative value that is alienable and applies to things that are fungible. I draw on this significant distinction throughout the discussion. My account is not Kantian, however, because I do not claim that the account of respect I describe reflects what Kant means by his concept. Moreover, my account is not Kantian because many readings of Kant suggest that he thinks that moral dignity is grounded in our capacities for autonomy. I do not follow this line of thought and instead I think moral dignity is a basic concept that applies to any being that meets the description of that kind of value, as I describe in Chapter 2. I do not think that it is necessary to be autonomous in order to have moral dignity, although being autonomous is sufficient for moral dignity.

For this reason, the account of respect that I describe is an egalitarian account that can apply to all humans directly, rather than merely derivatively or as an extension or "quasi" application of the concept. Respect involves a recognition and affirmation of the equal moral value of the other, but one need not have equal capacity, status, agency, or authority in order to have equal moral value. Although respect is a reciprocal-interactive second-person engagement, it need not be a symmetrical engagement. Symmetrical engagements call for a reversal of positions, or an imaginative projection into the position of the other. I draw on Iris Marion Young's (1997) description of asymmetrical reciprocity to describe the kind of relations involved in moral respect. Although I discuss asymmetrical reciprocity in the context of non-autonomous persons, Young developed the account when thinking of autonomous adults from different social backgrounds. I think asymmetrical reciprocity better describes respectful relations among both autonomous and non-autonomous persons. Asymmetrical reciprocity says that one cannot imaginatively project into the perspective of an other because we are non-interchangeable and our differences of experiences cannot be grasped by mere imagination alone. Young recommends that rather than imagining ourselves in another's place we ought to listen to the other's description of her perspective. I broaden Young's account to include other forms of communication beyond merely listening, for example we might also pay attention to body language.

1.5   Advantages

One advantage of the account of respect that I describe throughout this book is that it can explain why moral respect is important in health care from both the perspectives of health care professionals and patients. Furthermore, it can do so without impugning the motives of any of these groups. Respect is important in health care contexts because medicine, medical epistemology, medical research, and the technocratic demands of bureaucratic medical intuitions all involve a pull toward a third-person objective stance. Medicine focuses on repairing and treating the object-body as a physiological object that is interchangeable with other objects suffering similar disease. This objective stance is useful for scientific advancements and tracking the use of health care resources, but unless it is balanced by respectful second-person engagements with patients they can be left dehumanized. Moral respect is particularly important in health care contexts because it recognizes patients as more than mere objects and affirms their equal moral value. From a patient's perspective, the humanizing recognition of moral respect can help them live well with illness and it can help heal the alienation one might experience as one's pained body is poked and prodded by medical tests and procedures.

A second important advantage of this account is that it can directly include people who are not autonomous. I do not posit some capacity or set of capacity (psychological or otherwise) to ground moral dignity. Instead, I think that the value described by moral dignity is a basic concept that attaches to the whole, complex, and multifaceted person. Since moral dignity is not grounded in some capacity it cannot be acquired or lost, which better reflects the inalienable nature of moral dignity. Further, because it is not grounded in some capacity that people could have or exercise to a greater or lesser extent this better reflects the absolute nature of moral dignity.[7]

---

[7]Although I do not have space to consider whether animals also have dignity in this book, it might turn out that some or even many do have dignity. I would be happy with this result.

1.6   Limitations

The account of moral respect that I describe in this book takes time. One cannot reduce respectful second-person interactions to a check list or a set of tasks. The very attempt to do so would be to change the activity into a third-person concern for meeting the requirements of the list or the task rather than a second-person responsive engagement with the other person. For this reason, the kind of respect I describe requires institutional support. It is not something that could easily be taken up by an individual health care professional in an efficiency maximizing institution. It is (relatively) quick to perform tasks on a body, but it takes time to engage in second-person moral respect in non-objectifying ways. Many health care institutions stress their commitment to respect for patients, but then leave this moral obligation to their staff. This would be inappropriate for the concept of moral respect I describe here. For example, in Tronto's (1993) discussion of caregiving she stresses the importance of the response to care and how this ought to inform the identification of care needs. Providing responsive care in this manner might require structural adjustment to hospital policies (depending on their current organizational structure). The concept of second-personal moral respect has advantages in explaining the role of the concept and in its egalitarian application. It is limited, however, because the time involved requires material and institutional support.

Throughout the book I assume that we all agree that equal moral respect is an important ideal and worthy goal. This view has deep roots in liberal philosophy, feminist philosophy, and democratic traditions. I do not offer any argument to support this assumption because when I began working on this book the value of equal moral respect seemed settled. This is, perhaps, no longer something we should assume. There might be social shifts back toward inegalitarian conceptions of appraisal respect (perhaps better termed 'esteem' or 'honour') on the basis of merit or social group membership. In the future I might find I have to argue in favour of this assumption, but that argument is beyond the scope of this book. Even if there are social shifts occurring that question the importance of equal moral respect, this ideal remains firmly in place in health care contexts and health care ethics, at least for now.

Since my audience for this book are those who work and study in these contexts I omit a defence of the importance of equal moral respect and rely only on agreement with my predecessors.

1.7 Conclusion

To sum up—the account of moral respect that I sketch in this book is grounded in dignity which I described as an absolute, intrinsic value that inheres in all non-fungible individuals equally. Because we have dignity we are owed moral respect. Nevertheless, our dignity is not the target of respect—it is not that we respect this type of value. Instead, the person themselves is the target of respect because dignity is a type of value that inheres in non-fungible entities who cannot be considered interchangeable. For this reason, the target of respect is the person themselves. Finally, respect is enacted on this account when we engage with another from an interactive second-person standpoint. From the second-person standpoint we recognize the other as more than a mere object and we affirm their equal moral value. We disrespect another when we engage with them merely or primarily in ways that objectify them, or reduce them to a mere object. Denying or violating a person's autonomy is one significant way to reduce a person to a mere object. This denial and violation applies only to those who have sufficiently capacity to make autonomous choices. If violations or denials of autonomy were the only way to objectify another then my account would be inegalitarian. However, autonomous and non-autonomous patients alike can be treated merely instrumentally in, for example, research contexts. For autonomous patients instrumentality is typically avoided by obtaining informed consent. The argument I present in this book is that although informed consent is important, it is not sufficient to avoid objectifying disrespect. In addition, we must also consider the other eight forms of objectification. These other eight forms of objectification are ways both autonomous and non-autonomous patients can be mistreated and reduced to a mere object.

Disrespect occurs when a patient is *reduced* to an object, and this aspect of *reduction* is central to my account. A patient is reduced to a mere object when they are primarily or solely treated as an object. Treating someone as an object along one dimension is not sufficient for disrespecting them so long as the overall context is respectful and does not reduce the person to a mere object along multiple dimensions. In addition, a *temporary* near total reduction to a mere object along several, or even most, dimensions, is not always sufficient to disrespect a person. In emergency situations, when care is urgent, many dimensions of objectification might be permissible or even required as part of what it means to recognize and affirm a patient's dignity. For example, a patient might require urgent surgery and there might be insufficient time to obtain a properly informed consent. Surgery would require treating this patient as violable and surgical skills require treating this patient as interchangeable with, or sufficiently similar to, previous patients with the same condition. If this patient is unconscious and during surgery it is important to treat them as inert, because at the time they are. Focusing on the urgent repair of the trauma might require reducing them to their body and the parts of the body that are the focus of the surgery. In such emergency case, an extreme form of objectification is permissible as long as it is temporary and does not continue to affect the relationship once the emergency has passed.

My account suggests there are multiple ways in which we can reduce another to a mere object which means there is some vagueness to the account. There is no clear-cut way to determine "how many dimensions" need to be considered or which are "most significant" in all cases. Even violations of autonomy, which are almost always impermissible, are sometimes permissible, for example when an autonomous patient poses a risk of harm to themselves or others. My account would suggest that in these cases additional attention should be paid to the other dimensions of objectification in order to balance out the violation along one of the dimensions. Although this account contains some vagueness, I believe it is appropriate to the subject matter.

1.7.1 Chapter Summaries

In Chapter 2, I draw on the distinction between the interactive and objective stances to explain why moral respect is a central concept in health care ethics. Medicine, medical research, and medical epistemology are inherently objectifying. Medicine focuses on treating the body as a physiological object and develops knowledge about generalized disease kinds. Objectification is sometimes morally permissible and other times morally troubling—a context of respect can help to distinguish between these situations. I defend the egalitarian nature of the account of dignity described in the book against a recent challenge that claims dignity is inherently inegalitarian. I suggest that dignity is a basic concept and the inegalitarian implications arise when we mistakenly think we need to ground dignity in some set of capacities.

In Chapter 3, I use the account of moral respect developed in the previous chapter to apply it in concrete cases of medical care for non-autonomous patients. I consider the importance of respecting the dignity of children to their development of self-respect and I draw on Dillon's concept of care-respect as a second-person concept. I consider care for patients with Alzheimer's disease and suggests that when patients cannot advocate for themselves there is a particular danger of objectification and so moral respect becomes all the more important. In such cases understanding second-person care-respect as asymmetrical can directly include non-autonomous persons among those who are owed respect because of their dignity. I describe a variety of ways that we can engage with others from the interactive stance and I pay particular attention to how these engagements can counteract the objectification present in medical encounters.

Chapter 4 describes phenomenological accounts of embodied agency and the objectification and alienation that are part of many experiences of illness. I argue that the account of respect that I describe would matter even to autonomous patients because it could help them deal with the phenomenological threat of serious illnesses. In health our bodies are the centre of our world but rarely the centre of our attention: there is a seamless unity between the object-body and the body as subject. Experiences of illness focus our attention on the object-body which is thematized as a problem and a limit to our agency. In disabilities present since birth this involves not a changed embodiment, but a lack of fit with the habitus of the social world. Phenomenological accounts of illness and disability explain the importance of moral respect in health care contexts from a patient's perspective. Recently several philosophers have argued that empathy should play a greater role in health care contexts. I argue that respect does a better job of dealing with objectification than empathy because respect recognizes the subjective experience of the patient but without inappropriately adding to the emotional labour of health care professionals.

In the conclusion I offer a summary of the egalitarian concept of second-person moral respect described in this book. Moral respect is grounded in dignity and affirms the equal moral value of each member of the moral community. The concept of moral respect is egalitarian because it is not grounded in contingent facts or capacities. Moral respect affirms a person's value as more than a mere object. I summarize the various forms of objectification described in the book. I return to the question of balance to consider criteria for determining when balancing objectification will be particularly important. I consider whether the concerns discussed in this book could be addressed by an expanded account of autonomy or a more robust process of informed consent and I offer some considerations against this suggestion. I consider the advantages and limitations of the account of second-person asymmetrical care-respect. This account has the advantage of explaining the importance of respect in health care contexts without presuming that healthcare professionals might be predisposed to exploit patients. Moral respect is important in health contexts because of the inherently objectifying nature of health care. The discussion of moral respect is limited because the relations involved take time and material resources; they cannot be reduced to a checklist or set of tasks.

References

Baylis, Françoise. 2017. Still Gloria: Personal identity and dementia. *International Journal of Feminist Approaches to Bioethics* 10 (1): 210–224.
Beauchamp, Tom, and James Childress. 1979. *Principles of biomedical ethics*, 1st ed. New York: Oxford University Press.

Beauchamp, Tom, and James Childress. 2009. *Principles of biomedical ethics*, 6th ed. New York: Oxford University Press.

Beauchamp, Tom, and James Childress. 2013. *Principles of biomedical ethics*, 7th ed. New York: Oxford University Press.

Darwall, Stephen. 2006. *The second-person standpoint*. Cambridge, MA: Harvard University Press.

Dillon, Robin. 1992a. Care and respect. In *Explorations in feminist ethics: Theory and practice*, ed. Eve Browning Cole and Susan Coultrap-McQuin, 69–81. Bloomington and Indianapolis: Indiana University Press.

Dillon, Robin. 1992b. Respect and care: Toward moral integration. *Canadian Journal of Philosophy* 22 (1): 105–132.

Dillon, Robin. 2010. Respect. In *The Stanford encyclopedia of philosophy*, ed. Edward N. Zalta, Fall 2010 edition. http://plato.stanford.edu/archives/fall2010/entries/respect/. Accessed 12 June 2012.

Johnson, Tara, and Meredith Schwartz. 2007. *Gestational trophoblastic neoplasia: A guide for women dealing with tumors of the placenta, such as choriocarcinoma, molar pregnancy and other forms of GTN*. Toronto: Your Health Press.

Kant, Immanuel. 1996. *The metaphysics of morals*, ed. Mary Gregor. Cambridge: Cambridge University Press.

Kitwood, Tom. 1990. The dialectic of dementia: With particular reference to Alzheimer's disease. *Aging and Society* 10 (2): 177–196.

Kitwood, Tom. 1993. Toward a theory of dementia care: The interpersonal process. *Aging and Society* 13 (1): 51–67.

Kitwood, Tom. 2011. Dementia reconsidered: The person comes first. In *Adult lives: A life course perspective*, ed. Jeanne Katz et al., 89–99. Bristol: Policy Press.

Kontos, Pia. 2005. Embodied selfhood in Alzheimer's disease. *Dementia* 4 (4): 553–570.

Langton, Rae. 2005. Feminism in philosophy. In *The Oxford handbook of contemporary philosophy*, ed. Frank Jackson and Michael Smith, 231–257. Oxford: Oxford University Press.

National Commission for the Protection of Human Subjects of Biomedical and Behavioral Research. 1979. *The Belmont report: Ethical principles and guidelines for the protection of human subjects of research*. http://ohsr.od.nih.gov/guidelines/belmont.html. Accessed 30 Jan 2010.

Nussbaum, Martha. 1995. Objectification. *Philosophy & Public Affairs* 24 (4): 249–291.

Strawson, Peter. 1974. *Freedom and resentment, and other essays*. London: Methuen.

Toombs, S. Kay. 1988. Illness and the paradigm of lived body. *Theoretical Medicine and Bioethics* 9 (2): 201–226.

Tronto, Joan. 1993. *Moral boundaries: A political argument for an ethic of care.* New York: Routledge.

Welsh, Moira. 2018. The fix: One Peel nursing home took a gamble on fun, life and love. The most dangerous story we can tell is how simple it was to change. *The Toronto Star*, June 20. http://projects.thestar.com/dementia-program/. Accessed 21 June 2018.

Young, Iris Marion. 1997. Asymmetrical reciprocity: On moral respect, wonder, and enlarged thought. *Constellations* 3 (3): 340–363.

2

Dignity, Respect, and Objectification

Abstract Schwartz draws on the distinction between the interactive and objective stances to explain why moral respect is a central concept in health care ethics. Medicine, medical research, and medical epistemology are inherently objectifying. Medicine focuses on treating the body as a physiological object and develops knowledge about generalized disease kinds. Objectification is sometimes morally permissible, and other times morally troubling—a context of respect can help to distinguish between these situations. Moral respect is important in health contexts because of the inherently objectifying nature of health care.

Keywords Moral respect · Dignity · Medical research · Clinical care · Respect for autonomy · Informed consent

2.1 Introduction

In this chapter I outline an account of moral respect that is egalitarian and applies to both autonomous and non-autonomous persons alike. In the next chapter I provide further details on how this account applies

© The Author(s) 2019   **23**
M. C. Schwartz, *Moral Respect, Objectification, and Health Care*,
https://doi.org/10.1007/978-3-030-02967-8_2

to non-autonomous patients in particular. I propose the central role of 'respect' is the recognition and appreciation of the equal moral value of all members of the moral community. I begin in Sect. 2.2 by describing how I understand the concept of dignity. Dignity names a particular kind of moral value that applies to those who are unique, irreplaceable and whose value is absolute and inalienable. Affirming a person's dignity involves recognizing they are more than mere objects and interacting with them as subjects of particular experiences, as persons who can, in turn, react to us. Thus, a respectful engagement with persons can help to counteract the objectification that is an inherent part of the experience of illness and encounters with medical bureaucracies. Moral respect affirms our dignity and renders the objectification inherent in medicine morally permissible.

Darwall's account of moral respect as a second-person concept can help us understand why it is that contexts of respectful relationships render the objectification inherent in medicine morally permissible. Darwall draws on Strawson's distinction between the interactive and objective stance. The objective stance is inherent to medicine, since the goal of medicine is to treat the physiological body. The objective stance adopts the perspective of a detached observer and considers the other as an object of treatment in various forms (Strawson 1974, p. 9, n. 10). Darwall develops this account to describe respect as a second-person concept and respectful relations engage with others occur from an interactive stance. The second-person view of 'respect' stresses the interactive stance in which we engage with others in ways that are not possible to engage mere objects. Darwall's account is limited for its application in medicine, however. First, it is inegalitarian (a problem I return to in Chapter 3). Second, the focus on reactive attitudes as a means of interacting from the second-person perspective does not fit easily within health care settings because reactive attitudes are seldom appropriate in medical contexts. I examine the second objection in this chapter. I suggest that by considering the fuller account of objectification described by Nussbaum and Langton we are able to broaden Darwall's account to consider ways of engaging with others from the interactive stance that go beyond reactive attitudes. By focusing on the role of respect in countering objectification, we open space for a more diverse account of a moral respect that can cover a greater number of patients, including those who are not autonomous, or cannot be rendered autonomous, and therefore are not covered by the principle of respect for autonomy (Beauchamp and Childress 2013, p. 108).

In this chapter I develop a multi-faceted approach to objectification drawing on Martha Nussbaum (1995) and Rae Langton (1992, 2005) and introduced in Chapter 1. Nussbaum suggests that "in all cases of objectification what is at issue is a question of treating one thing as another: One is treating *as an object* what is not really an object, what is, in fact, a human being" (1995, pp. 256–257; italics in original). Typically the concept of 'objectification' is most frequently invoked to describe the sexual objectification of women, which is the context Nussbaum and Langton are writing about. But there is no reason that objectification must be restricted to that context because one can treat a person as an object in a number of different contexts. Objectification reduces the individual's body to a thing, or treats a person as fungible, or an instance of a generalizable kind in a way that downplays the person's subjective experiences and unique individuality to an impermissible extent. Considering a broader array of the ways objectification functions in medical contexts is important for understanding the role of respect in these contexts. Further, at least some forms of objectification are inherent to medicine, as I describe in this chapter, and hence are ineliminable. Other forms of objectification are external to medical practice, but might bear importantly on health care professionals' attempts to form respectful relationships with their patients. If the reason respect is important in medical contexts is to counter the inherent objectification present in medical practice, then the fullness of respect will require considering these other forms of objectification as well.

Moral respect is well-placed to counter the inherently objectifying aspects of medical encounters because respectfully engaging with others involves relating with them as more than mere objects among objects. The objective stance adopts the perspective of a third-person, detached observer, external to the relation. In contrast, respect involves a second-person relation of engagement and interaction with another. One reason that respectful relationships are well-suited to countering the objectifying elements of medicine and illness is that in respecting another we recognize and affirm the other's dignity. Respect, as an interpersonal engagement, recognizes that other as unique and non-fungible: no one but that particular person could be in this particular relation. Respectful relationships involve affirming subjectivity, non-violability and so forth. Attending to a variety of forms of objectification, and considering moral respect that directs us to affirm a person's dignity allows us to include non-autonomous patients among those who are directly owed respect. Although non-autonomous patients cannot have their autonomy violated, as they are not autonomous, they can be objectified in a variety of other ways and hence reduced to a mere object, as I explore in Chapter 3.

Before I turn to an examination of respect for non-autonomous patients I consider a challenge offered by Andrea Sangiovanni (2017) who claims that dignity is an inegalitarian concept. While I agree with Sangiovanni's criticisms of many conceptions of dignity I believe the mistake in these accounts and in Sangiovanni's discussion is the reductive attempt to ground dignity in some singular mental or psychological capacity. Instead, dignity attaches to whole, multifaceted selves. I elaborate this claim in Chapter 3.

2.2   Moral Dignity

The concept of dignity, like the concept of respect, is both widely used in bioethics and criticized for its lack of clarity (Macklin 2003; Beauchamp and Childress 2009, 2013). Ruth Macklin has suggested that the term 'dignity' could be eliminated from bioethics without "loss of content," since she believes, the concept reduces to respect for autonomy or respect for persons (2003, p. 1420). Although I agree that 'dignity' is often applied uncritically, I do not think the meaning of moral dignity reduces to that of *autonomy* or *persons*.[1] First, *moral dignity*

---

[1]Throughout the book when I use the term 'dignity' this should be understood to mean "moral dignity." There are other non-moral uses of the term, for example, one might say one lost one's dignity, where this means something akin to the idea that one was embarrassed. The focus of the book is on moral dignity, but I will often use 'dignity' alone to cover that concept. If I intend the

is a concept that describes an absolute moral value that is intrinsic to each person and applies equally to all, whereas *autonomy* is something that people can exercise to a greater or lesser extent. Second, *moral dignity* identifies a kind of value and this value is a *property of individuals*, unlike their autonomy which is a property of decision-making (Mason and O'Neill 2007, p. 17). Finally, *person* is a status term that indicates one is a member of a moral community, whereas *moral dignity* is a property term that applies to the kind of value that inheres in persons. To be sure, the terms dignity, persons, and autonomy, are related but each term is distinct from the others and cannot be reduced to the others without loss of content. To help clarify the concept of moral dignity it is worth retuning to Immanuel Kant's discussion in *Foundations of the Metaphysics of Morals* and *The Metaphysics of Morals*.

In describing his conception of dignity, Kant draws a distinction between different kinds of value. One kind of value, which Kant calls *price*, is a value that we assign to those things that are useful to us. Things that have market value can command either a higher or lower price depending on their usefulness, rarity, and the like. *Price* is an extrinsic value and one person might be able to command a higher price than another depending on their talents or skills (1995, pp. 51–52; AK 4:434–435). Further, the same person might be able to command different prices at different times throughout their life as they develop new skills, or the market values their existing skills differently. So price does not represent the kind of value that recognizes human equality. Price is a value of exchange that applies to commodities and to things that are fungible and can be traded one for another of the same kind without loss of value. In contrast, persons have dignity, which is an absolute intrinsic worth that does not admit of degrees. Kant writes that as a person one "is not to be valued merely as a means to the ends of others or even to his own ends, but as an end in himself, that is he possesses a *dignity* (an absolute inner worth) by which he exacts *respect* for

---

term 'dignity' to refer to some other form (e.g. "the dignity of the Prime Minister's office") I will specify its non-moral meaning.

himself from all other rational beings in the world" (1996, p. 186; AK 6:434–435; emphasis in original). Notice that here Kant grounds respect in our dignity, saying that dignity is what allows us to exact respect from others. The kind of value that dignity represents recognizes the non-interchangeability, or non-fungibility, of persons. As a kind of value, dignity puts persons outside of (and above) systems of market exchange that involve worth only in the sense of price. Dignity is a value term that describes the absolute, inalienable moral worth of persons, a kind of value beyond the values of exchange. Because this moral value is absolute, it does not admit of degrees, one cannot have dignity to a greater or lesser extent, and so all persons have equal dignity or equal moral value. In light of this value we are owed respect which recognizes and responds to the dignity in each of us. Since dignity describes a kind of value, this makes the concept of dignity different from the concept of autonomy because autonomy does not describe a kind of value; instead it describes a way of deliberating. Further, although Kant says that persons have dignity, dignity is a property of persons and not reducible to the concept of persons or their personhood.

I believe Kant's account of dignity provides a good basis for saying why we must respect one another. To summarize, according to Kant, we must respect one another because we have a certain kind of value, dignity, and this value makes us non-interchangeable with others. As a concept, dignity is both universal and particular. It is universal because each member of our moral community has an absolute moral value, which is equally shared with all others. But part of the important understanding of this kind of value involves understanding that we are not interchangeable with one another. It is the uniqueness of our relations, perspectives, and experience of the world, or our particularity, that makes us non-fungible. Although dignity forms the basis of respect, understanding dignity should point to a different target of respect. It is not that dignity *itself* is what we respect (its target), instead it is the *person themselves* that should be respected. So on this account the target of respect is the person themselves, and not merely some abstract fact or feature about the person such as their dignity. To understand why the person themselves is the target of respect, I believe it is useful to turn to Stephen Darwall's (2006) discussion of 'respect' as a second-person concept.

2.3 First-Person and Second-Person Perspectives

Darwall (1977) made a distinction between recognition respect and appraisal respect to help us understand the kind of respect Kant thinks is owed equally to all given that we also think that some forms of respect can be merited or not. Darwall suggests that appraisal respect involves the judgment that our behaviour or projects are worthy of praise, since not all projects and behaviours are equally respect-worthy, appraisal respect is not owed equally to all. In contrast, recognition respect does not involve an assessment; instead, it involves recognizing something as the kind of thing that elicits our respect and according that thing weight in our deliberations. Since recognition respect does not involve an evaluation but instead involves recognition it does not admit of degrees and can be owed equally to all. According to Darwall, recognition respect involves "a disposition to weigh appropriately in one's deliberations some feature of the thing in question and to act accordingly" (1977, p. 38). Moral recognition respect involves the added element that failing to accord the fact or feature the appropriate weight would be morally wrong. In his later work, Darwall (2006) continued to invoke the appraisal-recognition distinction, but he came to view his earlier discussion of recognition respect as mistaken because the target of recognition respect in the early account was a fact or feature of the person, rather than the person themselves. For this reason, Darwall began to describe respect as a second-person concept.[2]

Darwall's idea of the second-person standpoint relies on distinctions among the different possible perspectives that moral agents can adopt when evaluating or deliberating about a given situation or action. A first-person perspective when deliberating about a possible action or

---

[2]Here it is worth noticing that the account of respect for autonomy provided by Beauchamp and Childress (2013) has a form similar to Darwall's early (1977) account of respect for persons. On Beauchamp and Childress' view, respect involves recognizing a fact or feature about persons (their capacity to make autonomous decisions) and health care providers are to weight that fact or feature appropriately in their deliberations and constrain their actions accordingly by seeking informed consent from the patient in question.

a past action is one that evaluates our own actions or character "from the inside." The first-person perspective involves the domain of goodness and concerns questions about what *I* should want and what is good for people. The first-person standpoint involves an individual's relations to themselves. From this standpoint a moral agent might deliberate about what *I* should want now, considering what *I* wanted in the past and what *I* might want in the future. For example, I might have an immediate desire to skip class, but I might also have a more general first-person desire to obtain an education. Deliberating from the first-person perspective I might decide that my desire to skip class is incompatible with my desire to obtain an education and so I might decide to attend class in order to fulfil my more long-term goals of obtaining an education. Ideals of the good help moral agents to take-up the standpoint of a prudent person with foresight who might be concerned to harmonize her current desires with one another and with the needs and interests of her future-self (Anderson 2005). The first-person perspective has not played an important role in discussions of respect for others, but this perspective might be significant when deliberating about the requirements of self-respect and our duties to ourselves.

The second-person perspective involves the domain of right and wrong and of justice. The second-person stance enables us to take-up the perspective of others who might make claims on us because of how our conduct affects their interests (Anderson 2005). From the second-person perspective we consider the other (the grammatical second person singular, or "you") to be the source of our obligations. From the second-person perspective I have obligations to you because you have the standing to make claims on me. This second-person perspective differs from the third-person perspective that relies on objective considerations as the source of our obligations. The third-person perspective is that of a detached observer who is not involved in the situation. This perspective is sometimes called the God's-eye view. It involves standards of virtue and vice. From the third-person perspective we consider how a detached observer would judge, approve or disapprove of our conduct or the conduct of others (Anderson 2005).

As Carla Bagnoli describes, part of what makes the third-person perspective of an external judge seem like a morally attractive perspective is the promise of objectivity and impartiality that it offers. She argues, however, that the second-person perspective is not a partial perspective: the second-person perspective recognizes the particularity of individuals and their interpretation of events, but is not a form of special pleading (Bagnoli 2007, p. 117). The second-person standpoint for reasoning about moral issues is one that is well-suited to the recognition of dignity since, like dignity, this standpoint can account both for the particularity of individuals and the importance of their perspectives and experiences but without losing sight of the universal and absolute value that unites us as equal members of the moral community. The second- and third-person perspectives are not mutually exclusive. Although it would be difficult to adopt both at the same time, we can move between the two perspectives when considering and reasoning about a single moral issue or possible course of action.

When applied to the concept of respect these perspectives would give us a slightly different understanding of the concept. The second-person and third-person views differ because the third-person view of moral deliberation is characterized by giving the proper weight to some fact or principle in one's deliberations, whereas second-person respect is characterized by a particular mode of engaging with others in one's interactions with them. This distinction is not meant to be too strict, since even the second-person respect will involve some deliberation and the third-person moral deliberations certainly do not prohibit second-person respectful engagement. Instead, the distinction is meant to mark the main perspective adopted by each form of moral deliberation. These three different perspectives are not mutually exclusive; we often move between the three different perspectives when assessing a single moral issue.

Respecting another, according to Darwall, requires taking a second-person stance toward that person. The second-person stance is a relational stance because it recognizes that persons have interests and make claims on one another and that the validity of these claims depends on the other's relation to us a person. Darwall's (2006) account of respect as second-person concept stresses the importance of viewing persons as a source of authority and a source of valid moral claims. Respectful relations among persons involve an interactive stance from which we account for ourselves and ask others to give account to us, making our reasons explicit to one another. On this view, respect is an interactive relationship among individuals that is quite different than the objective stance we might take when deliberating from a third-person perspective or when considering another as a thing. Respectful second-person relations involve a mutual vulnerability where we recognize the other as an equal member of the moral community.

Darwall's discussion of second-person respect is useful because it stresses the relational nature of respect and helps us understand why when we respect another we are respecting *them* and not merely some fact or feature about them: The second-person standpoint involves a way of being *with* another. This engagement with the other recognizes that we can participate in activities with others who respond to us in ways that we cannot participate in activities with things that cannot respond (Langton 1992, p. 487). When we treat others as mere things we ignore their dignity and the particularity of their perspective. In contrast, from the second-person perspective we do things *with* another, and not merely *on*, *to*, or *for* the other. Darwall's account of respect as a second-person concept is also limited, however. First, because his focus is equal authority and autonomy, his account cannot deal well with respect that one might show to a child or someone who is no longer autonomous. Theoretically Darwall's account of 'respect' remains inegalitarian, as not all people can be self-originating sources of demands issued with equal moral authority. This makes the development of self-respect into a mystery, as I describe in Chapter 3. Second, Darwall focuses on reactive attitudes, and as a practical matter it would often be inappropriate for health care professionals to blame or praise patients, or ask them to account for their actions. Practically, it is not clear that reactive attitudes are appropriate in health care contexts. Although health care providers have their own values and might in fact think that patients are responsible for their disease, this is usually not appropriate to express in health care contexts.

This aspect of Darwall's account makes it an uneasy fit with health care contexts where reactive attitudes have only a limited role. Although it might sometimes be appropriate for health care providers to ask patients to account for their behaviour, for example because it might give clues toward a diagnosis, reactive attitudes have a limited role in health care contexts. I have argued elsewhere that assigning responsibility to patients for their illnesses or even maintaining their health can create climates of distrust in health care especially when patients lack resources to meet these responsibilities (Schwartz 2009). In health care contexts, health professionals are frequently taught that they should identify their values and bracket these values when treating patients. This is because blaming patients or asking them to account for their behaviour can easily shift into inappropriate paternalism as a result of the power imbalance inherent in the relationship. Although there is room to ask patients about their goals and values, the power imbalance in healthcare settings means that praise and blame can quickly undermine relationships of trust. Patients who are overweight, for example, report that too frequently when they go to a medical appointment it ends up in a lecture about losing weight. This can discourage people from seeking medical advice and undermine the professional disinterest that is the positive aspect of an objective stance in health care settings. Fortunately, the interactive stance might involve reactive attitudes to a certain extent, but as I explain in the next section interacting with someone as more than a mere object goes beyond reactive attitudes. I have already briefly suggested that caring for others from a second-person rather than third-person perspective is one way to treat them as equally morally valuable. I return to a discussion of care in Chapter 3.

To summarize the account of moral respect that I have given so far, I have suggested that moral respect is grounded in moral dignity. Dignity is a description of a kind of moral value that is absolute and so inheres equally in all persons. Dignity is a concept that points both to a universal moral equality and to the particular, non-fungible person. Respect is an interactive engagement with others that attends to their reactions and seeks to understand their perspectives and experiences, in whatever way these can be communicated. The interactive stance is a way of treating persons as more than mere objects; we can interact with persons in ways we cannot interact with objects and so a different kind of engagement is appropriate. Respect, in Robin Dillon's words, "has the resources to maintain a constructive tension between regarding each person as *just as valuable* as every other person and regarding this individual as *special*" (1992, p. 75; emphasis in original). Next I turn to a discussion of what it means to treat persons as mere objects.

2.4   Treating Persons as Objects

Engaging with others from the interactive second-person stance and not merely from the objective third-person stance is one way to ensure that we treat others as more than a mere object. The important element here is the idea of being *reduced to* a *mere* object. It is morally permissible to treat others as objects, as long as we treat them as more than a *mere* object at the same time. In this section I draw on Nussbaum, Langton and Toombs' accounts of objectification to provide an overview of what it means to treat someone as an object. In addition to treating others as a mere object, it is possible to objectify ourselves. I return to self-objectification in more detail in Chapter 4.

In health care ethics, the understanding of treating someone as a mere object has focused on two forms of objectification: *instrumentality* and *denials or violations of autonomy*. We treat someone instrumentally when the objectifier treats the person as a tool or mere means to his or her ends that are not shared with the person who is objectified (Nussbaum 1995, p. 257). Health care ethicists have described the danger of treating someone instrumentally, particularly as it applies in research settings. When a researcher is pursuing scientific knowledge, especially when that knowledge has the potential to benefit a great number of future patients, then there is a danger that the researcher might treat the research subjects as a mere tool to be used in order to create new knowledge. In clinical contexts the danger of denying or violating autonomy is well recognized and protected through obligations to obtain informed consent about medical treatments. One denies the autonomy of another when one treats someone who is autonomous as though they lack the capacity for self-determination (Nussbaum 1995, p. 257). In addition to denying that someone has capacities for self-determination, one might recognize these capacities and yet violate that self-determination by specifically ignoring the expressed wishes of the individual (Langton 2005, p. 246). Violations of autonomy are part of the sadistic pleasure someone might get from exercising power and control over the will of another person. Since both *instrumentality* and *denials or violations of autonomy* have been thoroughly discussed in health care ethics, they will not be my focus in this book, though I will discuss them when they are relevant. Instead, I focus on some forms of objectification that have not received as much attention from health care ethicists, but that are important to consider in medical contexts.

A third form of objectification occurs when a person is treated as *inert* in the sense that they lack agency or activity (Nussbaum 1995, p. 257). Objects are inert, whereas people have agency and typically engage in spontaneous activities. In some cases in health care contexts, it will be appropriate to treat a patient as inert. For example, if a patient is in a coma then treating the person as inert might be an important part of respecting them through providing care. Patients in a coma lack the agency to feed or clean themselves and so providing proper care might require a recognition of inertness. This would not involve reducing a patient to a *mere* object because the reason to feed them and keep them alive affirms their dignity, even if it might involve treating them in part as an object of treatment who lacks agency. In contrast, one can engage in behaviours that do reduce the patient to a mere object, particularly if one treats a patient who has agency as though this were entirely lacking. An example of when a patient would be reduced to an inert object might be when their physical disability such as an inability to walk is taken to mean that they cannot speak for themselves and so all questions are addressed to their caregiver referring only to the person in the wheelchair as "her" in the third-person (Toombs 1988, p. 209). In such a case the physical disability is inappropriately thought to affect all agency and the person is treated inappropriately as a mere object.

When Kant describes the difference between dignity and price, he notes that things that are assigned a mere price are fungible, whereas persons are non-fungible (1995, p. 50; AK 4:434). The fourth form of objectification is *fungibility* in which the objectifier treats the object as interchangeable with other objects of the same type, or objects of a different type, without loss of value (Nussbaum 1995, p. 257). For example, one cord of wood of a particular grade can be exchanged with any other cord of wood without loss of value because cords of wood are fungible. Additionally, any given cord of wood of a particular quality could be exchanged for an equivalent sum of money. In contrast, as I described above, persons are not fungible and cannot be exchanged one for another without loss of value. Medical treatment will require treating patients as fungible to a certain extent. If one patient with a particular diagnosis was not interchangeable with another to a certain extent, then treatment decisions would be impossible. There would be no way to compare two patients with the same condition. However, this does not mean that patients with the same diagnosis should be treated as entirely interchangeable because of their condition. Part of Tara's complaint, from *Case B* described in Chapter 1, was that the doctor was treating her like any choriocarcinoma patient and he launched into a discussion of the importance of using birth control without taking into account that her situation was unique and he had performed a hysterectomy that would make pregnancy impossible. In this case she was reduced to a fungible choriocarcinoma patient because the doctor did not take her circumstances into account. The doctor also failed to ask Tara why she was not taking birth control, which meant he denied her agency and autonomy as well. The gynaecological oncologist further failed to engage with Tara's unique particularity by overlooking the fact that she would not need to be on birth control to prevent pregnancy because she had a hysterectomy, *one that he had performed*. From Tara's perspective the hysterectomy was a momentous and life-change event. Prior to the hysterectomy Tara wanted a large family and she was in the midst of being forced to revise her plans. The doctor provided information that "the" choriocarcinoma patient needed to know, but in doing so Tara was forced to sit through a discussion that only highlighted what she had lost as a result of the cancer.

A fifth form of objectification occurs when we treat something as *violable*, as lacking boundary-integrity (Nussbaum 1995, p. 257). Many objects are violable; for example, when I drop a piece of chalk and it breaks I have not done something morally impermissible. Persons, on the other hand have boundary-integrity and most of the time if we violate these boundaries we have done something seriously morally impermissible. In medical contexts, however, health care professionals are routinely called upon to peer beyond the boundary of the skin in a number of ways. Drawing blood for a test or to transfuse into another patient is a form of boundary violation. Many medical imaging tests are designed to allow the health care professional to peer beyond the skin and deep into the inner parts of the body. One reason that respect is so important in medical contexts is because this kind of boundary violation is inherent to scientific medical care. The mere fact that bodily boundaries are routinely violated does not mean that the person has been reduced to something violable; when a patient has given consent to the test, surgery, or image then they have not been treated as a *mere* object, even if the health care professional goes beyond the boundaries of the body. Here we must exercise caution, however, because consenting to one form of boundary-crossing does not entail that other forms are also thereby permissible. Cases in which a patient consents to surgery do not also give doctors permission to use their bodies to demonstrate how to perform gynaecological exams. To treat the body as violable in ways that were not consented to reduces a patient to a mere object even when consent has been given to treat the body as violable in other ways.

Treating someone as though they can be *owned* by another or bought and sold is a sixth way to treat someone as an object (Nussbaum 1995, p. 257). It might seem like this kind of objectification would be rare in health care contexts since patients are not usually seen as the property of health care professionals. However, cases like the HeLa cancer cells involve treating parts of patients as though they can be owned. One interpretation of the problem in HeLa cases is that the researchers involved failed to inform Henrietta Lacks that they would be using her cells and so did not receive her permission. Although the lack of consent is one problem with these cases, Jenny Reardon and Kim TallBear (2012) show that in the case of the use of Havasupai DNA it was not a mere disagreement about lack of consent, additionally there was a disagreement about how things ought to be valued. Researchers valued Havasupai DNA and the HeLa cancer cells as mere objects of study and market exchange. The researchers turned these body parts into a mere object of ownership, which in addition to violating informed consent desecrated the body parts.

A seventh form of objectification occurs when someone's *subjectivity is denied or violated*. Denials of subjectivity occur when a person is treated as though they lack experiences and feelings and subjectivity is violated when experiences and feelings are acknowledged, but treated as though they need not be taken into account (Nussbaum 1995, p. 257). For example, in *Case D*, Olive was treated as though her lack of self-determination means her experiences don't matter, and so she was objectified in this way. In *Case A* described in Chapter 1, the interaction was disrespectful because the obstetrician was failing to attend to both the mother's and the daughter's subjectivity. He did not take into account that as a mother of a daughter with Down syndrome she loved her child. He failed to attend to the way in which the mother was becoming increasingly protective of her daughter who began the interaction playing on the floor and finished the interaction on her mother's lap being tightly embraced with her mother's hands shielding her ears. Furthermore, he treated the daughter as though she was not there and couldn't hear the discussion that was taking place between himself and her mother. I think it was unlikely that the daughter understood anything that was being discussed, nevertheless she would have an experience of this interaction and the mother's increasing protection of the daughter was a signal that something had gone amiss.

When a person is *reduced to her body* or its parts or she is identified with her body primarily, this is an eighth form of objectification, according to Rae Langton (2005, pp. 246–247). Langton's focus is on sexual objectification, but a similar form of objectification can occur in medical contexts when a patient is treated as nothing more that the body that needs curing. In medical contexts this form of objectification can include being reduced to a diagnostic category. If a patient is reduced to "the schizophrenic" then this fails to treat that patient as anything other than the diagnosis.[3] In *Case A*, part of the problem with

---

[3] I am indebted to Tim Mt. Pleasant for his constant vigilance pointing to this dehumanizing language. Any mistakes in language are my own.

the interaction was that the obstetrician objectified the daughter by reducing foetuses with Down syndrome to "risks" that had to be managed. He acted as though one merely needed to know a diagnosis and this would provide all of the relevant information about that individual. That he did so in front of someone who had the diagnosis made the interaction one that reduced the daughter to an object rather than treating her as having dignity.

An additional way patients can be reduced to their bodies in medical contexts occurs when they are reduced to a "unit of work" or a task to be completed rather than also as a person who can interact with the health care professional. For example, nurses often must complete an inventory questionnaire as they check on their patients' well-being. This is not objectifying in itself, but it can become objectifying if the nurse is not allowed time to respond to a patient's interjections because the "efficiency" demands of the institution do not account for interpersonal interactions (Austin 2012, p. 36). In *Case C*, what made me feel disrespected was the lack of attention to my experience of the test and its after effects. The clinic clearly expected that some women feel faint after the HSG procedure; otherwise there would not have been several chaise lounges in the change room area. No one checked on me, however, which made me feel worthless; especially after the embarrassing experience of crawling across the floor in nothing but a medical gown. I felt like the clinic was regarding me as a "unit of work" from which to extract money, and once the procedure was over they were on to the clean-up so they could see the next "unit of work" as quickly as possible. The fertility specialist had just performed an intimate procedure, during which she had inserted a tube into my uterus to fill it with liquid, yet there was no recognition of my experience, which made me feel violated. I did not feel that my autonomy or self-determination had been denied. In fact, in the moment when I felt most disrespected I was "free to go," but that was precisely the problem. There was an odd tension between a room set-up for fainting and a lack of engagement with an individual patient who was feeling faint. I felt reduced to an object, or source of income. I thought silently to myself about the fertility specialist as I waited to the taxi to pick me up, "wow, I am not making a baby with you!"

A ninth form of objectification occurs when an individual is *silenced* or treated as though they lack the capacity to speak (Langton 2005, p. 247).[4] In health care interactions that have gone awry patients sometimes complain that they are not "allowed" to speak—or that the health professional is not really listening to them. Instead, they are merely listening for the symptoms. In *Case B* one reason Tara felt disrespected was that she found the interaction silencing. Whereas she was usually open and outgoing, during this exchanged she had to bite her tongue, as she told me after the appointment. The doctor was focused on ensuring that information about preventing pregnancy was conveyed and understood. His discussion was rigid and focused on what "choriocarcinoma patients" need to know, rather than on what Tara, a patient with a hysterectomy, needed to know. As a result, Tara felt the informed consent discussion reduced her to a mere object, rather than engaging with her respectfully as a unique patient. Shortly after this happened Tara changed gynaecological oncologists.

The tenth, and final, form of objectification that I consider is *alienation*. It occurs primarily through a patient's own experience of illness and the way that the body, which is always present but usually forgotten, might come to be perceived as an obstacle to one's projects. For example, part of the experience of infertility in my *Case C*, was an experience of my body as an obstacle rather than a seamless vehicle for meeting my plans. For Tara, her hysterectomy meant she had to rethink her previous plans for a large family. Objectification of one's own body is characteristic of first-person phenomenological experiences of illness. Encounters with health care professionals can intensify the experience of alienation as we recognize our bodies as an object of the medical gaze (Toombs 1988, pp. 215–216). Undergoing the tests to see whether I had blocked fallopian tubes I could feel parts of my body for the first

---

[4]Langton considers *reduction to appearance* or treating another primarily as they appear to the senses. Her discussion focuses on sexual objectification, and this aspect of objectification seems more relevant in that domain and less relevant in health care contexts. I will not deal with *reduction to appearance* in this book. The types of objectification I describe here are not meant to be exhaustive, instead the types I identify are at least some of the forms of objectification that arise in health care contexts.

time, which was an uncanny experience. In Tara's *Case B*, she came to describe her well-being in terms of objective measures of her tumour markers rather than through subjective reports. Bodily alienation is the topic of Chapter 4—where I explore phenomenological accounts of illness as objectifying.

In this section I described a variety of ways that a person can be objectified. Although instrumentality and denials or violations of autonomy are important ways a person can be reduced to a mere object, additional forms of objectification are also possible. Attending to these additional forms of objectification provides a broader array of second-person responses that respectfully engage with patients in clinical settings where reactive attitudes are not always appropriate.

2.5 The Objective and Interactive Stances in Medicine

So far I have claimed that respect is grounded in our equal moral value and that the interactive second-person stance of respect is a recognition of that value. Interacting with an other involves recognizing that persons can participate with us in ways that mere objects cannot. So second-person respect treats others as more than a mere object. I have also considered a variety of ways we can objectify others and I have described how some forms of objectification are unproblematic when balanced by an interactive engagement of second-person respect. When not so balanced the person is reduced to a *mere* object.

In this section, I describe why a broader consideration of objectification is important in medical contexts in particular. My claim is that biomedicine involves a focus on the object body and a pull toward the objective, third-person stance. Hence there is an inherent tendency for health care professionals to objectify patients. Only in considering the varieties of objectification can we consider the importance of respectful second-person relations in counteracting this objectification. The primary goals of medicine include preventing or curing illness or treating its symptoms when cure is not possible. Medicine seeks to relieve pain and suffering and to promote health and longevity. The primary goal of medical research is to find new ways to achieve the goals of medicine. To achieve these goals, biomedicine and research considers the body as a physiological and biochemical object that can be manipulated and compared to what we know about other bodies as physiological objects. Hence, the goals of medicine inherently involve a pull to what Strawson calls the objective stance.[5] The objective stance is a perspective we adopt when persons are considered as "an object of social policy; as a subject for what, in a wide range of sense might be called treatment; [someone] to be managed or handled or cured or trained" (Strawson 1974, p. 9, n. 10). I believe the objective attitude plays an important and inherent role in medicine, which has treatment or cure as a goal. Strawson contrasts this attitude with an attitude of involvement, participation, or interaction in human relationship. He notes that while the two attitudes are "profoundly *opposed* to each other" they are not "*exclusive* of each other" (1974, p. 9; italics in original). That is, we can adopt both attitudes towards one another, perhaps shifting between them.

One reason that respect is important to consider in medical contexts is that the medical perspective involves taking an objective (third-person), rather than interactive (second-person), stance toward the patient. When a physician contemplates possible treatments for a given condition there is a pull toward thinking of the person primarily as a body or as locus of physiological processes. When we consider persons from this perspective we are less interested in the person as a subject with experiences and agency. We are instead looking at the person as a 'thing' that we might study as part of the natural order. Medical knowledge is primarily about the physiology, chemistry, and biology of persons and is less about their agency or understanding their reasons for action. Adopting the perspective of one who wants to treat another characterizes what Strawson calls the objective stance (Strawson 1974, p. 9, n. 10). Taking a medical perspective often involves (at least in part) seeing the patient's body and the patient's symptoms as an interchangeable

---

[5]I am indebted to Langton's (1992) discussion in her paper "Duty and Desolation" for pointing out the importance of this passage in Strawson.

instance of other similar symptoms that have already been studied by medical science, and can therefore be treated or managed. From a medical perspective there is a sense in which one body is interchangeable for another body; the body is treated as a fungible thing.[6] The creation of and reliance on impartial scientific evidence inherently involves treating an individual as a physiological object of study or diagnosis to be manipulated and cured.

Darwall focuses his account of second-person respect on reactive attitudes and recognizing the other as a source of self-originating claims and reasons, but he pays less attention to the role of second-person respect in countering a potentially dehumanizing objectification that considers the other merely as an object. Since the object-body becomes a central focus in illness, I believe that health care ethics should consider this latter role for second-person respect in medical contexts. While I think Darwall's account of the role of reactive attitudes in the second-person standpoint is important, these attitudes probably have a lesser role in medical contexts, as I described in Sect. 2.3. One reason that objective attitudes in medical contexts might be useful would be to lessen the chance that doctors or nurses engage in reactive attitudes that might blame the patient.

Nussbaum has demonstrated that objectification is not always morally problematic when it occurs within a context of respect and regard for humanity (1995, p. 289). In medical contexts some objectification of the patient is likely desirable, since it is difficult to see how the knowledge gained through scientific research could apply to a patient if we did not see patients as fungible to some extent. We need to see a patient as relatively interchangeable with other patients experiencing the same illness or medical knowledge gained from studies performed on others and clinical expertise gained from interacting with others would never apply to them. However, when objectivity is out of balance, and there are inadequate second-person respectful interactions, then a moral

---

[6]This point might be contentious, since some doctors try to treat the "whole patient" and would resist the description that they are treating only the patient's body. We are able to move between perspectives that we take on one another, however. My point here is merely that some of the scientific methods involved in medicine encourage a focus on the body.

problem of disrespect arises. Not only are we treating a patient as an object, were are treating them as *nothing more* than an object. Moral respect is important in health care practice because it reminds us that although treatment might involve some objectification we must guard against allowing this objectification to reduce a patient to a *mere* object. Darwall's description of 'respect' as a second-person concept that involves an interactive perspective is useful for describing respectful relations as engaging with others in ways that go beyond how we can engage with mere objects. Engaging respectfully *with* another acknowledges that the other can participate in activities and respond to our mutual participation. In contrast, objects cannot participate, respond, or engage in mutual activities (Langton 1992, p. 487).

Respectful second-person engagement requires a nuanced balance among perspectives. This is similar to the importance of balance in the atmosphere. CO_2 is not 'bad' in itself, and it might even be desirable or necessary for some things we think are good. Nevertheless if CO_2 is out of balance with the other elements of the atmosphere this can be bad for the stability of the climate and the things that were once nourished by the CO_2 might come to be in trouble. I think Darwall is right to stress the importance of second-person engagements, but I am not convinced that these are best demonstrated through reactive attitudes. For this reason, I consider a broader range of forms of objectification. By thinking about additional forms of objectification we open space for a variety of forms of second-person interactive engagement with others in ways that oppose the ten forms of objectification I have described. Second-person interactive engagements will be particularly important when a number of these dimensions of objectification are present at the same time, especially when a person's first-person experience of alienation is particularly acute.

For example, medical encounters typically involve a focus on the object body from an objective stance that seeks to treat or cure some physiological process. This might also involve tests and imaging that challenge the inviolability of the body's boundaries. If the illness is serious a patient might come to feel alienated from her own body. In this kind of situation reactive attitudes alone are unlikely to counter the profound objectification. By attending to the variety of forms of objectification we can also respond from the interactive stance by responding to the subjective experiences and by creating opportunities for patients to explain their first-person views that go beyond mere reports of signs and symptoms. We can treat someone from the objective stance while avoiding reducing the person to a *mere* object when we balance the objectification with an engagement from an interactive stance that recognizes and affirms the equal moral value of *this* patient who is unique and not fungible.

The idea of balance also explains why in some medical encounters informed consent alone will suffice, while in others something more will be needed. In many routine medical encounters the chances of being reduced to a mere object are negligible. For routine check-ups or getting a flu shot, any objectification that might occur is only minor and temporary. It is unlikely to lead to a deep sense of alienation from one's own body. In these cases, very little will be required to counteract the objectification involved in medicine. For a flu shot a brief discussion and request for consent is enough. In other cases, however, the experience of the illness will be more intense and the forms of objectification involved in the medical encounter will be more pervasive. These experiences of illness might lead to self-objectification as I describe in Chapter 4. In such cases to balance the objectification will require greater attention to the ten dimensions of objectification I have described.

Disrespect occurs when a patient is *reduced* to an object, and this aspect of *reduction* is central to my account. A patient is reduced to a mere object when they are primarily or solely treated as an object. To objectify a person along a single (or even a few) dimensions is not itself disrespectful as long as one is interacting with the other as a person with dignity along the other dimensions. The interactive aspect is important because it is also possible to be neutral with respect to one of the forms of objectification. They will not always be applicable in every situation.

To return to *Case B*, Tara's doctor was not treating her instrumentally, nor was he treating her as owned. Neither of these potential forms of objectification arose in that interaction. Although he had treated her as violable during the hysterectomy this was permissible because Tara consented. During the interaction described, the doctor was not treating her as violable, though he had forgotten he so treated her in the past. Finally, the doctor was trying to support Tara's autonomy by providing valuable (though unneeded) information about avoiding pregnancy. Some forms of objectification were irrelevant to the case, while respect for autonomy was explicitly upheld. In this case attention to balancing the objectifying elements of medical care could be expected to be particularly important because of the connection to alienation; serious illnesses, like cancer, are more likely to lead to situations of existentially unbearable forms of alienation from one's body. Serious illness brings one's mortality into focus. In addition, the life-saving surgery Tara's doctor performed contributed to her pervasive experience of her body as a mere object. In this case a respectful engagement that affirmed her subjectivity and agency could have countered the objectification that was causing significant changes to her life. This respectful engagement was not forthcoming, however. Instead of affirming her equal moral value, Tara experienced the doctor as reducing her to a mere object along a number of dimensions of objectification. He was treating her as interchangeable with other choriocarcinoma patients by providing her information relevant to the generic patient but not to her. He ignored or demeaned her subjectivity, and perhaps her agency, by failing to ask about the consequences of the hysterectomy he performed on Tara's plans for maternity. He did not pause to ask why she was not using birth control—instead, he treated the lecture as a task to "get through." Tara's changed demeanour and her increasing distress—which was readily apparent to me—did not affect the lecture. Finally, Tara experienced this as silencing—the doctor did not ask questions or treat her capacity to speak as significant. Tara responded by shutting down. Whereas she would typically speak up for herself the encounter reduced her to a "nothing," as she described it. I return to a consideration of the phenomenological experience of illness in Chapter 4. Considering a patient's experience of illness can help us identify when balancing objectification will be particularly important. In Chapter 3, I consider a variety of forms that a respectful second-person engagement can take. Before turning to these discussions I want to first consider a challenge to the egalitarian account of dignity I described in Sect. 2.2.

2.6   Egalitarian Moral Dignity

In Sect. 2.2, I suggested that moral respect, whose role is the recognition of the equal moral value of each person, is grounded in dignity and has as its target the person themselves. I described dignity as a property of persons that describes their intrinsic and absolute moral value and recognizes the unique non-fungible nature of each of us. Thus, dignity is both universal and particular. It is universal because all persons have equal dignity, but it is particular because as unique and non-fungible beings we are each non-interchangeable with other persons. *Dignity*, as I understand it, describes *equal* moral value and applies to all persons. As a kind of value, 'dignity' is contrasted with 'price,' which is the value involved in market exchanges. Dignity is an inner worth that is inherent in persons and not alienable. Dignity involves intrinsic value (something that is valuable for its own sake) rather than the mere instrumental value of price (something that is valuable as a means of reaching some further goal). Dignity applies to things that are non-fungible and cannot be interchanged one for another whereas price applies to fungible commodities. Dignity is an absolute value that does not admit of degrees; hence, the value of dignity is equal in all who have it, whereas something with price can be more or less expensive.

At this point, I need to defend the claim that dignity applies equally to all persons from the charge made by Andrea Sangiovanni in her provocative book, *Humanity without Dignity* (2017). Unlike the objection of vagueness raised by Macklin or Beauchamp and Childress, Sangiovanni believes that any "evocative and powerful" value will have a variety of meanings and uses that reflect the historical development of the concept (2017, p. 14). Her objection to the concept of dignity is that it cannot meet the two desiderata: (1) *equality*, which would explain the sense in which we are all equal in dignity; and (2) *rationale*, which explains why and in virtue of what we have dignity (Sangiovanni 2017, p. 15). Sangiovanni surveys three major traditions of dignity (Aristocratic dignity, Christian dignity, and Kantian dignity) and finds that each is unable to meet these desiderata.[7] If Sangiovanni is right, then it might seem that my account is not egalitarian and cannot apply to non-autonomous patients and therefore would fail to meet one of the desiderata that I laid out in the introduction. Namely, it would fail to be an egalitarian conception of moral respect. I begin by briefly explaining Sangiovanni's objection and I reply by disagreeing with her second desiderata: rationale. In particular, I think dignity is a property of persons that is basic and applies to our whole, multifaceted and complicated selves. I think that it is a mistake to think that there is some psychological capacity or set of capacities that gives humans unconditional worth. Dignity does not attach to psychological capacities alone but instead it attaches to our whole unique and multifaceted selves. Further, when Sangiovanni describes a being with, what she calls, 'basic moral standing' she invokes a description that sounds very much like what I have called 'dignity.' Hence, we need a conception of dignity to describe that particular type of value.

Sangiovanni's discussion of Kantian dignity is far more sophisticated than I have space to describe here. Briefly, Sangiovanni considers the 'Regress reading' that understands rational nature as having a special kind of value because as evaluative beings we are the source of all value in the world and must possess a different kind of value from the things in the world that we value. She also considers the 'Address reading' which holds that valuing, justifying, and other moralizing activity necessarily presupposes the "equal and reciprocal *authority*" of those we address through that activity (Sangiovanni 2017, p. 36).

---

[7]Sangiovanni describes Aristocratic dignity as applying to elevated social roles that fulfil worthy social goals and those persons who fill these roles (2017, pp. 16–18). The Aristocratic sense of dignity is not egalitarian as it does not attach to persons merely in virtue of our humanity. Instead, it attaches to an elevated role or position that involves an inequality in status, which would not give everyone an equal right to have their dignity respected (Sangiovanni 2017, pp. 19–20, 26). The Christian tradition, according to Sangiovanni, grounds dignity in man's reasoning capacities as a manifestation of his reflection of the image of God that sets him apart from all other creatures (2017, p. 28). This account of dignity cannot explain, or provide a rationale for the kind of value that humans are said to have without an appeal to the image and likeness of God (2017, pp. 31–32). This account lacks a philosophical basis that could be generally acceptable and so fails to meet the rationale desideratum. These are not the conceptions of dignity that I have invoked, so I do not look at these discussions in detail.

Although Sangiovanni's criticisms of the 'Regress reading' are original and, to my mind, persuasive, I focus here on the 'Address reading' since this is the one that captures the idea of second-person address described by Stephen Darwall. Sangiovanni suggests that the 'Address reading' does not reference the value of our capacity for rational choice, but instead our equal *authority* to make claims on one another.

Sangiovanni suggests that Darwall understands dignity as a kind of moral standing in terms of the authority to make demands on others and we are equal in dignity because we have an equal authority to make demands on one another (2017, p. 51). For this to hold, Sangiovanni claims Darwall would have to show that all second-person address presupposes equal authority and equal moral standing (2017, p. 52). However, Sangiovanni claims that second-person address does not (logically) necessarily make such presuppositions. We could coherently imagine the Christian God issuing second-person calls to account to His subjects. In such a case, God would not have to presuppose that those subjects have equal authority to hold Him accountable (Sangiovanni 2017, p. 55). One need not believe that the Christian God actually exists. Sangiovanni's thought experiment is sufficient to show that the *logic* of second-person address does *not* presuppose equal moral standing as Darwall claims. So Darwall's use of second-person address is not sufficient to show that we are moral equals or have equal dignity, understood as equal moral status (Sangiovanni 2017, p. 56).

I think Sangiovanni's criticisms of Darwall are persuasive. I do not think the account of respect Darwall offers is an egalitarian account precisely because his focus is on equal authority, which we do not all share. In addition, as I discuss in the next chapter, Darwall does not think we can have equal second-person moral respect for children to whom we show only "quasi-respect" according to Darwall (2006). However, I do not think the account of dignity and second-person moral respect that I have offered falls prey to this criticism. Notice that Sangiovanni understands 'dignity' to refer to equal moral *status*, where moral status involves equal authority to make demands. This is not the account of dignity that I have offered. I do not think that dignity refers to equal moral *status*. Instead, I have suggested that 'dignity' names a kind of *value*: an intrinsic, inalienable, absolute, unconditional, value that recognizes the non-fungible nature of its bearers. We do not need to have equal authority in order to have equal value. Instead, we need to fit the description of that kind of value, and if we do meet that description then we have dignity—which is the name for that value.

Do we need a concept of dignity to describe this kind of value? I think we do. The kind of value named by dignity differs from the kind of value involved in market exchanges, which Kant calls 'price.' Particularly in current societies where market thinking predominates and markets encroach on ever more aspects of social organization (Sandel 2012) it is important to consider that there are other non-market ways of valuing—and this is precisely what dignity describes. Furthermore, Sangiovanni invokes something that sounds very similar to the conception of dignity that I have described when she offers her own account of basic moral status. Sangiovanni suggests that basic moral status gives a claim to be treated only in ways that can be justified from a common perspective whereas equal moral status gives us the right to be treated as an equal (2017, p. 61). She argues that a creature with a conscious life that allows an evaluative point of view on the world is owed basic moral status. Such a creature matters morally "in its own right (i.e., not merely instrumentally) but also for its own sake (i.e., from its own point of view," she writes (2017, p. 65). If "mattering morally" is akin to having moral value then the non-instrumental, intrinsic value named by 'dignity' seems to be invoked in the description of creatures with basic moral standing described by Sangiovanni.

We need a concept of 'dignity' because it names a unique kind of value (Buckley 2013). It does not reduce to autonomy because autonomy is a property of decision-making whereas dignity inheres in the person themselves. Dignity describes equal moral value and so I agree with the first desiderata provided by Sangiovanni. I disagree with her rationale that requires a psychological capacity or set of capacities to ground dignity; that is, to explain why we have dignity. Instead, 'dignity' is explained by its description as an intrinsic, inalienable, absolute, unconditional, value that recognizes the non-fungible nature of its bearers, and any being that meets that description has dignity.

Second, I object to the reductive idea that only psychological capacities are candidates for identifying dignity or that we should presume there will be a single capacity or singular set of capacities that we would use to identify dignity. Humans are complex and multifaceted beings. To think that our dignity needs to be based on a single facet of our selves is to reduce that complexity. Instead, we have dignity because we have the kind of value described by the concept. We have dignity not in virtue of a single aspect of our self, but because of the fullness and richness of our selves. Indeed, it is this full richness that makes us non-fungible creatures with absolute intrinsic value. What then makes a human self? We are embodied, thinking, experiencing, emotional and social creatures. We grow up in particular relationships with others and our societies. We tell stories about our self understanding and our understanding of others. We have embodied agency, which involves the material aspects of ourselves through which we are recognized by others (Kontos 2005). I turn to a discussion of respect, dignity and non-autonomous persons in the next chapter.

2.7   Conclusion

To sum up—the account of moral respect that I have sketched in this chapter is grounded in dignity which I have described as an absolute, intrinsic value that inheres in all individuals equally. I return to a discussion of dignity in Chapter 4. Because we have dignity we are owed moral respect. Nevertheless, our dignity is not the target of respect—it is not that we respect this type of value. Instead, the person themselves is the target of respect because dignity is a type of value that inheres in non-fungible entities who cannot be considered interchangeable. For this reason, the target of respect is the person themselves. Finally, respect is enacted on this account when we engage with another from an interactive second-person standpoint. From the second-person standpoint we recognize the other as more than a mere object. We disrespect another when we engage with them merely or primarily in ways that objectify them, or reduce them to a mere object. Denying or violating a person's autonomy is one significant way to reduce a person to a mere object. This denial and violation applies only to those who have sufficiently capacity to decide. If this were the only way to objectify another then my account would be inegalitarian. However, autonomous and non-autonomous patients alike can be treated merely instrumentally in, for example, research contexts. For autonomous patients instrumentality is typically avoided by obtaining informed consent. The argument I present in this book is that this is necessary but not always sufficient to avoid objectifying disrespect. In addition, we must often consider the other eight forms of objectification. These other eight forms of objectification are ways both autonomous and non-autonomous patients can be mistreated and reduced to a mere object. In the next chapter I look at forms of respectful interactive engagements that go beyond reactive attitudes and include caregiving and a variety of means to counter the ten forms of objectification.

References

Anderson, Elizabeth. 2005. Dewey's moral philosophy. In *The Stanford encyclopedia of philosophy*, ed. Edward N. Zalta, Fall 2008 edition. http://plato.stanford.edu/entries/dewey-moral/. Accessed 14 Sept 2009.

Austin, Wendy. 2012. Moral distress and the contemporary plight of health professionals. *HEC Forum* 24 (1): 27–38.

Bagnoli, Carla. 2007. Respect and membership in the moral community. *Ethical Theory and Moral Practice* 10 (2): 113–128.

Beauchamp, Tom, and James Childress. 2009. *Principles of biomedical ethics*, 6th ed. New York: Oxford University Press.

Beauchamp, Tom, and James Childress. 2013. *Principles of biomedical ethics*, 7th ed. New York: Oxford University Press.

Buckley, Shannon. 2013. Why bioethics needs a theory of dignity. Presentation at the Canadian Bioethics Society conference. Banff, Alberta.

Darwall, Stephen. 1977. Two kinds of respect. *Ethics* 88 (1): 36–49.

Darwall, Stephen. 2006. *The second-person standpoint.* Cambridge, MA: Harvard University Press.

Dillon, Robin. 1992. Care and respect. In *Explorations in feminist ethics: Theory and practice*, ed. Eve Browning Cole and Susan Coultrap-McQuin, 69–81. Bloomington and Indianapolis: Indiana University Press.

Kant, Immanuel. 1995. *Foundations of the metaphysics of morals*, trans. Lewis White Beck. Upper Saddle River: Prentice Hall.

Kant, Immanuel. 1996. *The metaphysics of morals*, ed. Mary Gregor. Cambridge: Cambridge University Press.

Kontos, Pia. 2005. Embodied selfhood in Alzheimer's disease. *Dementia* 4 (4): 553–570.

Langton, Rae. 1992. Duty and desolation. *Philosophy* 67: 481–505.

Langton, Rae. 2005. Feminism in philosophy. In *The Oxford handbook of contemporary philosophy*, ed. Frank Jackson and Michael Smith, 231–257. Oxford: Oxford University Press.

Macklin, Ruth. 2003. Dignity is a useless concept: It means no more than respect for persons or their autonomy. *British Medical Journal* 327 (7429): 1419–1420.

Mason, Neil, and Onora O'Neill. 2007. *Rethinking informed consent in bioethics*. New York: Cambridge University Press.

Nussbaum, Martha. 1995. Objectification. *Philosophy & Public Affairs* 24 (4): 249–291.

Reardon, Jenny, and Kim TallBear. 2012. 'Your DNA is *our* history' genomics, anthropology, and the construction of whiteness as property. *Current Anthropology* 53 (S5): S233–S245.

Sandel, Michael. 2012. *What money can't buy: The moral limits of markets.* New York: Farrar, Straus and Giroux.

Sangiovanni, Andrea. 2017. *Humanity without dignity: Moral equality, respect, and human rights.* Cambridge, MA: Harvard University Press.

Schwartz, Meredith. 2009. Trust and responsibility in health policy. *International Journal of Feminist Approaches to Bioethics: IJFAB* 2 (2): 116–133.

Strawson, Peter. 1974. *Freedom and resentment, and other essays.* London: Methuen.

Toombs, S. Kay. 1988. Illness and the paradigm of lived body. *Theoretical Medicine and Bioethics* 9 (2): 201–226.

3

Respect and Non-autonomous Patients

Abstract Schwartz uses the account of moral respect developed in the previous chapters to apply it in concrete cases of medical care for non-autonomous patients. She considers the importance of respecting the dignity of children to their development of self-respect and draws on Dillon's concept of care-respect as a second-person concept. She considers care for patients with Alzheimer's disease and suggests that when patients cannot advocate for themselves there is a particular danger of objectification and so moral respect becomes all the more important. In such cases understanding second-person care-respect as asymmetrical can directly include non-autonomous persons among those who are owed respect because of their dignity.

Keywords Moral dignity · Non-autonomous patients · Children · Alzheimer's disease · Down syndrome · Care-respect · Self-respect · Care ethics

© The Author(s) 2019
M. C. Schwartz, *Moral Respect, Objectification, and Health Care*,
https://doi.org/10.1007/978-3-030-02967-8_3

3.1   Introduction

In this chapter I consider whether the kind of moral respect I have described can apply to non-autonomous patients directly rather than derivatively or by according them merely "quasi-respect." I argue that we can directly respect infants, people with serious cognitive disabilities and persons who have lost the cognitive capacities required for full moral agency and complex medical decision-making. Even non-autonomous persons and those who lack full moral agency have equal moral value. They have equal moral value because they have dignity—they meet the description of value that is named by this term.

In addition, I describe a diverse account of the moral domain; that is, what we ought to count as moral behaviour and practices. Many philosophers who have described moral respect draw on a Kantian account of morality. Kantians focus on moral duties and rights, the exchange of moral reasons, and holding one another to account for our moral decisions. I believe these are important elements of the moral domain but I do not believe they exhaust the kinds of behaviour that counts as moral. In addition to these moral behaviours we also offer care and comfort and attend to the moral emotions of those in our moral community. We attempt to repair moral transgressions and create worlds that are morally habitable for as many people as possible. We raise children to become moral agents with well-developed moral sensibilities. We tell stories that teach moral lessons about how to live well with each other and we tell stories to awaken each other to perspectives that differ from our own in an attempt to enlarge our thought and understanding of those whose lives, circumstances, values, and perspectives are quite different from our own. We develop narratives to explain who we are and whom we understand others to be. Moral reasons, accountability, and responsibility are important aspects of moral behaviour but they are not exhaustive of the ways we recognize the value of those we hold in our moral communities. Because moral activities comprise a broad range, many people with diverse capacities can participate in these activities, though they may not be able to participate in all of them.

Finally, in this chapter I return to the idea that respecting one another requires a balance. Respectful engagement with others recognizes their equal moral value and treats them as more than a mere object. I have been considering a number of ways in which we can objectify others. In Chapter 2, I claimed that objectification along one or even several dimensions is not sufficient to *reduce* a person to a *mere* object so long as one engages respectfully along the other dimensions. In this chapter I consider people who have not yet developed full moral autonomy or have lost many (or even most) of the capacities required for autonomy. To engage respectfully in these cases would involve attending to the other dimensions that make us more than mere objects, such as our subjective experiences and emotions, our embodied agency, and non-fungibility.

I focus here on children and patients with Alzheimer's disease due to the limitations of space. Everything I discuss about equal respect for non-autonomous persons would apply equally to persons with serious cognitive disabilities who were never and will never become autonomous. These considerations apply to those with serious cognitive disabilities because respect on my account does not aim at autonomy. The argument that we ought to respect children does use developmental concerns, but the argument is not that we should respect children in virtue of their developing autonomy. Further, the argument about respecting patients with Alzheimer's disease does not depend on their past autonomy that they once had and now have lost.

The central role played by the concept of 'respect,' which differentiates it from other similar moral concepts, involves developing and fostering relationships that respond to and affirm the equal moral value of another with whom we interact. When we act respectfully toward another we interact with them as more than a mere object among objects. Engaging with another respectfully inherently involves responding to that other, recognizing the other's own particular response to us and being attentive to these responses and affirming their importance. Moral respect is the moral concept that takes this equal moral worth as its main concern.

3.2   Respecting Infants and Children

In this section I consider whether the account of moral respect can apply to non-autonomous persons. I focus here on infants and children. I have drawn from Darwall the idea that moral respect is a second-person relation that engages with the other as a source of moral concern and recognizes their dignity—which I understand as a kind of *value* as opposed to equal moral status or equal moral authority. Darwall does not think his account can be applied directly to young children as they lack the equal authority he believes is presupposed by successful second-person address. Darwall believes we offer children "quasi-respect" but not actual moral respect. I think Darwall is wrong to think we cannot show moral respect to children. Darwall is correct that infants and children lack equal moral authority, and that their moral agency is still developing. However, this does not entail that they have lesser moral value. If moral respect recognizes this equal value then it is intelligible to suggest that we have and can show equal moral respect for infants and children.

According to Darwall, we do not show true respect to a child or an infant. Because, on Darwall's view, respect involves recognizing *equal authority*, and children do not have equal authority, recognition respect cannot apply to children in a straightforward way. This is a problem for Darwall's account because it is unclear how a child would ever develop a sense of self-respect without the experience of being respected by others. Further, if Darwall's account of respect is supposed to recognize the standing that members of the moral community have to make claims on one another (2006, p. 17) his account seems incomplete if it can only countenance the kinds of claims made by adults. Infants and children *do* make claims on us even before they can do so by offering explicit reasons of their own. Even new born infants make claims on us to care for them, and if we fail to do so we wrong *the infant* and not merely others in our society. The account of dignity that I offered in Chapter 2 can make sense of these kinds of claims. Once we recognize an infant has dignity and is a member of the moral community we have obligations to the infant, including those that recognize the infant has both equal value and uniqueness that is worth protecting. This recognition also involves a recognition of her claims on us for the care required to develop that uniqueness.

There are practical reasons to think we must be able to engage with infants and children from a respectful interactive stance. In particular, if we cannot show true respect to infants and children then the development of self-respect becomes mysterious. In order to develop a robust sense of self-respect we need to have the experience of being respected by others. Practical concerns related to the development of self-respect and moral agency suggest that being respected is crucial for that very development. However, the form that this respectful second-person engagement takes will differ in emphasis depending on the development or maturity of each person. I consider a broader range of moral response to evince this respect than does Darwall. It will often be inappropriate to hold young children to the same standards of accountability as we would hold most adults. There are, however, more ways to recognize the equal moral value of children (and adults, too, for that matter) than asking for reasons or meeting their behaviour with reactive attitudes. In this section I focus on just one of these ways of recognizing children's equal moral value: when we offer them second-person care-respect. Not all care is from a second-person perspective, some care can be given paternalistically. When we do provide second-person care, or what Robin Dillon (1992a, b) calls 'care-respect,' then this is a way to recognize the unique, non-fungible moral value of children. I begin with the practical concern about the development of self-respect and then turn to a consideration of caregiving in which we engage children from a second-person perspective.

As a preliminary consideration it is worth noting that descriptively we do engage in practices that respect people with a broad range of capacities. Current social relational practices do recognize the dignity of many members of our moral community, including those who are not autonomous and those whom we might not regard as moral agents. For example, when an infant is born, parents typically already recognize the uniqueness of the child and would believe that exchanging this child for another would entail a loss of value. The parents recognize the child has the kind of absolute, non-fungible value described by the concept of dignity. This is not a mere matter of valuing biological ties, because adoptive parents, too, would feel violated if their adoptive child were taken away and replaced with another. All of these exchanges would violate our recognition of the dignity of the individual. In other cases, for example those involving people with advanced dementia, we often continue to recognize an individual's value even long after they have lost the complex capacities for full moral agency. In these cases, we engage in holding them in personhood, to use Hilde Lindemann's term (2002), because we continue to recognize the person has dignity.

Not only do we recognize and affirm the equal moral value of infants and children, but it is also developmentally important that we do so. Without the experience of being respected it is mystifying to understand how children would develop self-respect. Robin Dillon describes how the development of basal self-respect requires the experience of being respected by others. If this respect from others is missing then basal self-respect can be damaged or distorted. I examine Dillon's account of the development of self-respect in detail. Following Dillon, my argument in this section is that developing self-respect requires the experience of being respected by others who are valued and valuable. Thus it must be possible to respect children or no one would develop self-respect. In order to understand the kind of respect that applies to children, we will need to consider second-person engagements that involve affirming the worth of the other, but are not always a reactive attitude. Developing self-respect requires the experience of being respected. Thus, it must be possible to respect children.

One important task of moral development involves gaining a secure sense of one's own equal moral worth—or self-respect. When one respects one's self, one has confidence in one's moral value and one's place in one's moral community. Unfortunately, not everyone develops this robust form of self-respect. Some people instead develop only a damaged or fragile sense of self-respect. Others might develop a sense of shame, inadequacy, or self-contempt. Robin Dillon (1997) suggests that which of these experiences we have depends on the intellectual, emotional and political experiences we have during our moral development.

Dillon's account of 'self-respect' understands it to have a complex structure (1997, p. 227). Philosophers have typically focused on the self-regarding beliefs, attitudes and judgements involved in self-respect and they have taken these cognitive evaluations to give rise to emotions such as pride or shame that are connected to self-respect but not essential to it (Dillon 1997, p. 228). Dillon disagrees with this account, which conceptualizes self-respect as a discrete entity. Dillon suggests instead that self-respect is a complex and multifaceted way of being in the world. At the core of this way of being is "a deep appreciation of one's morally significant worth" (Dillon 1997, p. 228). Because Dillon's account of self-respect focuses on a recognition of one's moral worth, or dignity, it coheres well as the self-regarding complement of the account or moral respect I have described last chapter.

Self-respect, according to Dillon, is a complex and layered experience that involves beliefs, emotions, and experiences. These are different layers, which can cohere or come apart. We might believe ourselves to be equally morally valuable and we might experience this value directly as a feeling that we are equally worthy (1997, pp. 228–232). When these aspects diverge we might believe ourselves to be equally morally valuable without being able to feel worthy. Dillon says there are different ways in which we can understand our equal moral worth: intellectually or experientially. We might understand this worth intellectually, in ways that can be explicitly articulated in words (perhaps with some effort), and the implications of these beliefs can be logically extended to novel contexts. For example, if we believe "all persons have dignity—that is equal moral value," we might then take this equal moral value to apply to ourselves as a logical extension of the original belief about all persons (Dillon 1997, p. 239). Intellectual understanding might engage one's emotions, though it need not do so. Experiential understanding involves feeling the truth of something directly and will typically engage our emotions. Experiential understanding need not engage our conscious beliefs or judgements (Dillon 1997, pp. 239–240).

Dillon suggests there are some things we understand adequately when we understand them intellectually. For example, if I understand a math theorem intellectually I do not need to additionally understand it experientially. For other things, the intellectual understanding by itself is defective unless it is accompanied by an experiential understanding (Dillon 1997, p. 239). Self-respect is the latter type. If we only hold correct beliefs about our equal moral value, but cannot experience ourselves as equally valuable—that is, we feel worthless even though we don't believe this is true—then our self-respect is damaged or incomplete.

The different ways of understanding self-respect lead Dillon to postulate the existence of what she calls 'basal self-respect' (1997, p. 241). Basal self-respect is an affective, implicit interpretive framework and orientation to the self that acts as a filter through which we interpret all other beliefs or indicators of our worth. Damage to basal self-respect can be particularly enduring because the shape that basal frameworks give to our self-experience is non-propositional, so it cannot easily be put into words (if at all), and is often invisible to one's explicit understanding; thus this damage is difficult to undo even when one holds beliefs that explicitly endorse one's own equality. The experience of being respected by others and growing up in a context that values human equality both implicitly and explicitly can lead to a robust basal confidence in one's equal moral worth.

Dillon's concept of basal self-respect shows why it is important to experience respect from others even before we have the cognitive capacity to develop *beliefs* about our own moral equality. Our ability to directly experience ourselves as equally morally valuable arises from our interactions with others who either interact with us *as* equally valuable or fail to do so. Some of this understanding might be explicit—something we are able to articulate in words—but, Dillon suggests, much of this understanding will be implicit and unable to be formulated as explicit beliefs. The implicit understanding acts more as a framework for interpreting the world than as something we can explicitly articulate (1997, pp. 240–241).

When we experience respect from others our basal self-respect involves a secure sense of our intrinsic and unconditional worth (or dignity). We experience ourselves as having moral value independently of our successes or failures. As Dillon writes, one develops "an abiding faith in the rightness of my being" (1997, p. 242). Unfortunately, basal self-respect can also be damaged, especially when our social context contains oppressive messages that some social groups are worth less than others. For example, Robin Dillon describes how in sexist societies women grow up in social contexts that are saturated with messages that "what is female is worth less" (1997, p. 246). Although these social contexts might involve explicit endorsement of human equality, they also involve a number of institutions and interactions that code women as less than fully equal. Dillon proposes that exposure to these messages, especially at a time before one can rationally engage with or criticize the devaluing messages, can lead to damaged basal self-respect.

Basal self-respect can similarly be damaged in racist societies that involve implicit messages that some races are inherently more valuable than others. For example, images and stories that depict African Americans as violent or dangerous erroneously encode messages about defective moral characters attributed to a whole racial group. Societies that have widespread prejudice against people with disabilities also send conflicting messages about the moral value of living with a disability (Scully 2008). These messages can be picked up prereflectively and damage basal self-respect.

Messages of lesser worth can be transmitted in various and often subtle ways. Dillon suggests we absorb and "metabolize" these messages in emotionally laden ways before we can cognitively interpret (and thereby criticize or resist) such experiences of being valued (or valueless) among a community of other persons who are "valued and valuable" (1997, p. 245). Although the development of basal self-respect arises through subtle means, its effects are powerful (Dillon 1997, p. 242). When basal self-respect is damaged the effects are wide spread and difficult to reverse. Basal self-respect involves a basic evaluative orientation toward one's own worth. This orientation is emotionally structured and as an experiential understanding much of this orientation lies beyond our explicit cognitive articulation. As a result it is hard to "get at" or change our basal self-orientations. When basal self-respect is damaged then we have a nagging sense, below the threshold of awareness, that colours our self-experience as one who is "not good enough" or who is unworthy (Dillon 1997, p. 242).

Dillon suggests that the importance of basal self-respect gives rise to political responsibilities. Our basal self-understanding develops in social and political contexts that influence self-valuations. Unconditional parental love can help us form strong basal self-respect. In oppressive social contexts, however, individuals will absorb messages of subordination and devaluation for categories of persons from the wider social world. These messages can hinder the development of basal self-respect despite a loving home life (Dillon 1997, p. 245). If Dillon's account of the development of basal self-respect is correct, then this gives us a practical and political reason to think that not only is it possible to engage respectfully with children in ways that affirm their equal moral value, but also that we have moral and developmental reasons we ought to do so.

So far I have suggested that the development of self-respect requires experiences of being respected by others who are themselves valued. This gives us a practical reason to think that it is possible to respect children, but I have not yet said what that respect would be like. In order to do so, I want to consider a broad range of moral practices, paying attention to caregiving as a form of moral response and moral address that are particularly important when interacting with children. In order to give a more detailed account of moral respect for children it is useful to return to examine some reasons Darwall offers against the intelligibility of full moral respect for children. I think Darwall is right when he says that respect requires engagement with the other from a second-person standpoint. But Darwall focuses too narrowly on the exchange of reasons, which neglects the relational aspects of moral agents. I begin by considering Darwall's account of caring for children, which he thinks is distinct from respecting them. I then develop a more nuanced account of caregiving inspired by feminist philosophers and care ethics.

Darwall (2006) characterizes respect as based on dignity, which he understands as equal status to make second-person claims. In contrast, he says, care is based on welfare. This creates a fundamental difference between care and respect on his view. Darwall sees care as third-person, welfare-regarding and agent-neutral; in contrast, respect is a second-person, dignity-regarding, agent-relative activity (2006, p. 126). This means that care takes promoting welfare as its central concern (rather than the person themselves) and welfare does not depend on the particular person cared for but is instead an objective criterion (third-person) that need not reference the cared-for in particular (agent-neutral). In contrast, respect recognizes the dignity of the person and so it takes the person respected as the central concern and a source of obligations (second-person) and it must reference the person or their values (agent-relative). The central concern of care-giving, according to Darwall, is whether it advances the cared-for's welfare which can be objectively determined (2006, p. 127). When we provide care to another we are not concerned with her particular values or perspectives as we would be in a second-person engagement. Instead, according to Darwall, care is guided by third-person concerns about what will objectively advance the person's well-being. For example, when a father refuses to give significant weight to his young daughter's protests about eating healthy food he is guided by concerns for her welfare and he takes her welfare to be the source of his obligations to provide healthy food. If the father considers his daughter's values this would only be to the extent that they affect her welfare; for example, he might be concerned that forcing his daughter to eat could result in issues with food later in her life, again a welfare concern. Darwall believes the father would be justified in pressuring his young daughter to eat healthy food since he is guided by concerns for her welfare. In contrast, if the father was now dealing with a daughter home from university, it would be disrespectful to force her to eat the healthy food he has provided because he would not be according her will the regulative weight it deserves (2006, p. 128). Darwall believes that this distinction places care and respect in different categories expressing different ways of treating people. I think that sometimes care is delivered in the third-person way that Darwall imagines. This is the kind of care that we call "paternalistic" care. Caregiving need not take this paternalistic form, however.

Many feminist interpretations of care have resisted formulations that conceptualize care as purely universal or third-person in Darwall's sense. A wealth of feminist theorizing about caring sees care precisely as involving attention to the particularity of the one cared-for and to the response one's care receives (Noddings 1984; Ruddick 1995; Tronto 1993; Dillon 1992a, b). Care ethicists typically conceive of caregiving so that it involves an engagement with the other as the *particular* person they are (Spelman 1978, p. 151). In each of these cases, care is described not as an agent-neutral, objective third-person attempt to promote welfare, but instead as an engagement and response to the particular person. These accounts do hold that caregiving aims at promoting welfare and meeting needs, as Darwall suggests, but they involve a second-person care because the person themselves, with their agent-relative needs and their particular response to caring activities is the source of caregiving obligations. According to Joan Tronto (1993), we cannot determine whether the caring need has been met *unless* we attend to *the response from those cared-for*. Robin Dillon characterizes this attentive valuing as care-respect (1992a, b). Care-respect is a second-person response to the concrete, particular person who reciprocates through responses to the care as it is received. Caregiving can take the paternal-istic form that Darwall describes, and when it does it would not be a way to respect another. The paternalistic care doesn't treat the other as a being with the universal absolute value of dignity that points to the concrete non-fungible particularity of a person. Instead, paternalistic care promotes the paternalist's idea of what is in someone's welfare inter-ests, perhaps by appeal to an objective standard. In contrast, care-respect uses focused attention to understand and value another on their own terms (Dillon 1992b). Tronto argues that conceptually care is both uni-versal and particular (1993, pp. 109–110). Care is universal because all humans have needs and require care from others, but it is also particu-lar because the shape of those needs will be particular to an individual and depend on biology, age, and the cultural forms that caregiving takes (Tronto 1993, p. 110).

Joan Tronto provides an explicitly political account of care ethics that considers how society provides (or fails to provide) care that does not presume caring involves only two people (such as a parent and child), nor does it presume that caregiving is a private activity which takes place only in the home. Tronto and Bernice Fisher offer a broad defi-nition of care as "a species activity that includes everything we do to maintain, continue, and repair our world so that we can live in it as well as possible" (quoted in Tronto 1993, p. 103). As an ongoing pro-cess they divide care into four phases that are analytically separate. First, *caring about* involves recognizing that care is necessary—noting both that there is a need and deciding it should be met. Second, *taking care of* involves taking responsibility for meeting a need and deciding how to respond to it (Tronto 1993, p. 106). Third, *care-giving* involves the direct, physical work of meeting the need. Finally, *care receiving* recog-nizes the response from the care receiver to the care provided (Tronto 1993, p. 107). Tronto suggests that ideally the four phases would be well-integrated but in reality there is often conflict among the four phases and the four phases may be divided up and distributed among different groups, structured by power. In bureaucracies, Tronto suggests, there is often distance between deciding which needs to meet and how, and the actual caregiving work and care receiving. As a result she sug-gests they may not provide very good care. Putting distance between the first two phases of care and the last two will also result in third-person care as I have been describing it.

According to Tronto, when we include the response to care as part of what it means to care well, then we can attend to certain dilemmas that arise in care-giving contexts. For example, the one providing care might prefer to do so in a different way than the care-receiver, but it is not immediately clear whose preference is more compelling (as it is on Darwall's view where the objective welfare interests take precedence when we are third-person caring on a paternalistic model rather than second-person care-respecting). Tronto argues that caring well requires attending to the response from the one cared-for, although the kinds of responses that we attend to will differ in different circumstances. Sometimes these responses will take the form of explicit reasons or expressed preferences, but other times we have to rely on other cues such as emotional expressions or perceived discomfort.

Darwall is right that not all caregiving is a form of respect but the kind of care described by care ethicists does meet the description of respect I have offered: it is grounded in dignity, recognizes the equal moral value of a person and treats them as more than a mere object. Care-respect responds to another as non-fungible as having unique and particular responses to the care received. When care-respect is offered to children this is a way to directly respect *them*. In Dillon's words, care-respect responds to the individual human 'me-ness' (1992b, p. 118). Sometimes care-respect will aim to consider or develop autonomy. Sara Ruddick suggests one of the virtues of mothers is learning to relish the uncontrollable will of their children (1995, p. 73). When care seeks to help children develop independence and autonomy it involves the virtues of letting go. But autonomy is not a prerequisite for this kind of care-respect. Dillon notes that autonomy is one path to becoming a 'me' but it is not necessary for a person to be autonomous in order for them to deserve respect (1992b, p. 119).

In sum, there are developmental reasons that we should consider respecting children to be not only possible, but morally required. Without experiences of being respected the development of basal self-respect becomes mystifying. One way to directly respect children involves engaging in second-person caregiving. Since caregiving is often sentimentalized in Eurocentric societies, and because it is often associated with parents (particularly mothers) caring for their children, there might seem to be little relevance between this discussion and the provision of health care. I turn to that consideration now to draw out the relevance for health care contexts.

First, although I have been discussing care as a way of respecting children and infants, care and the need to be cared for is not restricted to the young. We all need care throughout our lives and so this discussion applies to adults as well as children. Second, recall in Chapter 2, I described a variety of ways in which medical care and medical knowledge inherently involves adopting a third-person perspective on patients in a way that focuses on their body and its physiological processes as an object for treatment. Now, using the distinction between paternalistic third-person care and second-person care-respect we can see that medicine additionally involves an allegedly objective determination of well-being, which it then promotes using treatments, prevention, or cures. Medicine describes a particular view of what constitutes 'health' and medical researchers attempt to find ways to restore or preserve 'health' on that understanding. So the care involved in health care can be third-person health-regarding care. It need not be only that kind, however. We can additionally provide care-respect in health care settings.

3.3   Returning to *Case A*: A Child with Down Syndrome

Recall from Chapter 1 I described a case where I was observing an ultrasound that was conducted in a high-risk pregnancy clinic at a major urban hospital.[1] A pregnant woman came in with her five-year-old daughter. Her daughter had translocation Down syndrome and genetic tests found that it had been inherited. Since this meant there was up to a 15% chance of having another child with Down syndrome, the woman's

---

[1]Some details of this case have been changed or omitted to protect the identities of those involved.

second pregnancy was being carefully monitored. As the mother sat on the ultrasound table, her daughter played happily in the room. The mother smiled at her daughter, and they laughed together as the technician prepared the equipment to perform the ultrasound.

After the ultrasound was complete, the doctor came into explain what the imaging technology had found. The nuchal translucency scan (NT scan) results were abnormal and showed that the fetus was at a greater "risk" of having Down syndrome.[2] The doctor began to explain the risks of different prenatal genetic tests, the attendant risk of miscarriage and the "risk" that the fetus had Down syndrome. He helped the mother assess the relative risk of miscarriage as it compared to the risk of Down syndrome and he explained all the options that were available to her. During this discussion, the mother stiffened and stopped smiling.

"My baby's fine," she said through pursed lips.

"That may be true," replied the doctor. "These tests are only screens and we won't know for sure whether the baby has Down syndrome until we confirm via genetic test."

Again the doctor explained the results of the NT scan, this time putting the "risk" ratios into context in a more detailed way. The mother lifted her daughter onto her lap and held her head as he spoke. He asked whether she wanted to book the genetic test with the receptionist on her way out.

The mother seemed uncomfortable, agitated and as if she wanted to leave. "I don't want a test," she said, "my baby is fine." She began to gather her things and said, "thank you doctor."

Once she left the room, the doctor commented to me, "she's in denial. I couldn't get through to her, she was being very difficult."

Recall that in this case the mother later described the doctor as disrespecting *her daughter*. Was the mother right to think so? Her daughter was likely too young to have an intellectual understanding of this exchange. She probably did not understand the relative risk

---

[2] I put "risk" in quotation marks to highlight that it is not entirely obvious that this is a risk. Labelling it as such presumes the condition is a 'harm' since risks are calculated using the magnitude of harm and its probability of occurring. Elsewhere I have written about prenatal genetic testing for "risks" and the effects this can have on social trust (Schwartz 2007).

information conveyed. The daughter was not the subject of medical decisions in this exchange as the mother was given information and asked to make decisions about her own pregnancy care. If we think of the principle of respect for autonomy understood as accepting patients' choices, it is difficult to interpret how this interaction might be considered disrespectful. The doctor did not violate the woman's autonomy. Quite the contrary, he was clearly explaining the results of the NT scan and putting the "risks" into context. When the woman's reaction that her baby was fine indicated to him that she had not understood the "risks" as he was presenting them he went over the information again, using more examples and contextual information each time. The doctor was very patient with her while she was in the room and he was doing everything he could to help the mother reach a reasonable, informed decision.

If we consider the discussion of basal self-respect this can help us understand the mother's judgement. The mother need not have an explicit theoretical understanding of basal self-respect to want to protect her daughter from what she perceived as negative messages about people with Down syndrome. The daughter need not be able to explicitly understand the technical medical discussion to "pick up" negative evaluations about members of her social group. Dillon suggests that such messages are often picked up at an implicit level below cognitive awareness. The mother's increasing protection of her daughter makes sense on this analysis. However, the increasing tension between the mother and the doctor might itself contribute to the implicit messages that devalue people with Down syndrome. Body language and emotionally laden exchanges are some ways in which such messages are encoded.

The way the doctor presented the information seemed to assume that the reasonable thing to do was for this woman to pursue further testing, and consider abortion if the tests were positive. His explanations implicitly conveyed that the risk ratios of the various tests, the risk of miscarriage, and the "risk" of the fetus being affected with Down syndrome could be used as an algorithm to objectively determine whether it was worth undergoing various tests. From his assessment and description of the risks, he clearly thought that further testing was called for. Since the "risk" that the fetus was affected with Down syndrome was quite high in this case, he seemed to think that considering the possibility of further testing was a reasonable next step despite the increased risk of miscarriage.

The information was presented in a neutral way that focused on the objective third-person "risks" discovered by the NT scan, but the possibilities that were considered and described focused on the medical interventions. He did not suggest, for example, that she might forego further testing and allow the pregnancy to progress without further intervention beyond routine prenatal care. That is, he did not suggest that she simply bring the fetus to term with Down syndrome. The focus was instead on the unusually high chance that the fetus had Down syndrome and the importance of confirming the diagnosis and considering further medical interventions. Nothing was explicitly said to devalue people with Down syndrome, but the way the options were constructed and the risks explained implicitly conveyed that it was bad to have the "abnormal" condition.

If we consider the discussion of care-respect and the four phases of caring we might identify additional problems in the interchange. The doctor's discussion seemed to presume that a healthy pregnancy was one in which the foetus did not have a genetically detectable difference. So the way the "need" for testing was presented included the determination that a thick NT scan result indicated a "risk." This might have been the way the doctor would identify a need, or it might have been decided by a manager of the hospital. Perhaps the doctor's response to the mother was limited by policies set by those deciding what to *care about* and how to *care for* some issue in a way that did not allow sufficient attention to the response to care in the moment. Improving relationships between health care providers and patients sometimes requires changing structural relations between *caregivers* and managers so that there is more room to include *responses to care* in the decisions about what requires care and how to care for those things.

In many ways the doctor was meeting his obligations. He was carefully explaining the information and he has a legal and moral obligation to do so. He was not, however, carefully attending to the response his caregiving was receiving from the mother. She was not receptive to the discussion and became increasingly agitated. This should have indicated that the care was not well-received and the need had perhaps been misidentified. Rather than engaging in second-person care-respect, the doctor seems engaged in third-person paternalistic care that uses the doctor's (or perhaps the medical institution's) account of the needs of pregnant women in general, thereby treating them as fungible.

How would second-person care-respect suggest the interaction could change? Much of the disrespect in this exchange could have been alleviated by simply asking the mother to identify her own need *before* the onset of the informed consent discussion. Although some bioethicists suggest there is little point to prenatal genetic testing *unless* one is willing to abort (Shenk 2000, p. 85), this is not actually the case. There are a number of reasons one might want testing other than possible termination. One might simply want information to be prepared. There are some complications associated with Down syndrome, such as heart malformations, that can be corrected with surgery. When I spoke with the mother later she said she was getting tested to be prepared in just such a circumstance. Had the doctor asked the mother what need she saw for the test then she might not have become so defensive and they would have been working together to meet the need rather than working at odds.

Perhaps if the doctor knew more about basal self-respect or had a more nuanced understanding of moral respect he might have paused before launching into the informed consent discussion to see whether they could talk in private while someone else watched the daughter for a minute. An institution responsive to patient needs might have a child care facility for such occasions. In this interchange the doctor focused on respecting the mother's autonomy understood as providing objective, full and accurate information about possible medical options. In the course of doing so, however, he reduced people with Down syndrome (and thereby the daughter) to a fungible disease type. He failed to attend to the mother's subjective reactions of distress and the daughter's subjective absorption of negative messages about people with Down syndrome. Although the doctor did a good job of explaining the options and screening results he objectified both the mother and daughter throughout the discussion.

3.4   Respecting Patients with Alzheimer's Disease

To this point I have been discussing moral respect for infants and children. Some of that discussion was developmental—I described the development of self-respect and second-person care-respect as engaging with another in a way that is attentive to their response to care. Some care-respect recognizes the dignity of children by helping them develop independence and the skills that are needed for full moral agency and autonomy. Hence, one misreading of my discussion would suggest that it is children's potential for autonomy that gives them dignity. That is not my position. For one thing, not all children will become autonomous. Some children have cognitive disabilities that preclude the complex reasoning described by autonomy. These children still have the equal moral value described by dignity—they are unique, non-fungible persons with particular subjective experiences and thereby deserve moral respect.

To further forestall the misperception that it is the potential for autonomy that confers dignity I turn, in this section, to a consideration of respecting people with a diagnosis of Alzheimer's disease or other forms of advanced dementia. Those who are experiencing memory loss and the loss of other cognitive functions often experience periods of lucidity as well as periods of confusion, anxiety, and difficulty with daily tasks. At some point, however, autonomy is lost and will not return. On the account I offered throughout this book, this will not entail a loss of dignity—understood as equal moral value.

I return to the idea of 'balance' which I introduced in Chapter 2 as a way to avoid reducing someone to a mere object that is especially important when one might be treating someone as an object along one or even several dimensions of objectification. If we are treating another as an object and, as I described in Chapter 2, medical care inherently involves treating the body as an object, then strengthening our attention to treat a person as more than a mere object along the other dimensions is a way to avoid *reducing* them to a *mere* object. In this section, I pay particular attention to fungibility, inertness, denial of subjectivity, silencing, and reduction to the body as particularly important, though somewhat complicated, dimensions of objectification when we are respecting people who are experiencing changes in their cognitive capacities.

Something is fungible when it can be exchanged with something of the same type or an equivalent type without loss of value. For example, a candy bar is fungible because it does not matter which candy bar you have, all of the same kind have interchangeable value. We can also exchange equivalent money for a candy bar. People are non-fungible because they cannot be interchanged without loss of value. Some discussion of Alzheimer's disease uses metaphors that encourage us to view people with Alzheimer's disease as fungible. The cognitive changes are sometimes described as a "loss of self" (Kontos 2005, p. 553). As Françoise Baylis noted about her experience with her own mother, people might invoke these metaphors as a way of reassuring family members. For example, when Baylis and her mother had an argument, friends might try to reassure her by saying "it isn't your mother [speaking or acting] it is the disease" (Baylis 2017, p. 220). Although the metaphors are evocative and meant to be reassuring, they represent the person as fungible: if we lose our self then we become an interchangeable body or an instance of a disease kind. Further, as Baylis describes, although these comments are meant to be reassuring, they rob the family member of the meaning of their experiences. The experience of having a parent lash out at you is emotionally difficult precisely because she is your mother and not merely some "demented woman" (Baylis 2017, p. 221).

Tom Kitwood (2011) suggests that these metaphors are not always invoked in a benign way. He says that they sometimes result from a defensive reaction to fear and anxiety about frailty, dependence, dying, and mental illness. For many people these anxieties manifest in a defensive tactic that turns people with a diagnosis of dementia into "a different species" or something not fully human (Kitwood 2011, p. 90). This gives rise to metaphors of dementia as "a death that leaves the body behind" (2011, p. 91), which again encourages us to view the person as interchangeable with any other body "ravaged by dementia." The very act of labelling someone with a stigmatized disease can encourage treating them as a "disease object, and outcast," according to Kitwood (1990, p. 183).

In addition to metaphors for Alzheimer's disease and defensive mechanisms that encourage the reduction of people with Alzheimer's disease to an interchangeable "body" or "disease kind" sometimes standards of care can contribute to this form of objectification as well. For example, until recently the regulations for long-term care in Ontario, Canada, were task-oriented, as was Olive's care described in *Case D*. Health care professionals and staff were required to undertake a demanding schedule of feeding, cleaning, and changing residents in the facilities. All of this work had to be documented and submitted to the government for tracking. The documentation was meant to ensure the safety of the residents and it was implemented in response to a previous audit that found some long-term care facilities were neglecting even these basic tasks because the residents were unable to complain. The tasks were, however, only allotted minimal time for completion in order to promote "efficiency" and protect the bottom line (Welsh 2018). Although the regulations were meant to prevent neglect, they encouraged staff to view their interactions with the residents as "a task to complete." Since the tasks were standardized and did not allow time for second-person respectful interactions this would add further incentive toward treating residents as though they were fungible tasks to "get through" as quickly as possible.

Because these metaphors, defensive mechanisms and task-oriented care requirements all tend toward treating people who are experiencing the various stages of dementia as fungible, this will be one dimension along which it is particularly important to resist—or balance against—this tendency. People do not become fungible because they receive a diagnosis of Alzheimer's disease. One might lose the cognitive capacity for complex autonomous choice but, as Dillon notes, autonomy is only one route to 'me-ness' (1992b, p. 119). Additionally, as I described in the discussion of dignity in Chapter 2, our relationships, our narratives, and our embodied agency make up part of our non-fungible 'me-ness' that make us intrinsically valuable. Part of what makes us who we are is our relationships to others as parent, friend, child, co-worker and so forth. We retain these important relational connections and they make us irreplaceable. Though, as Françoise Baylis notes, these relational aspects of the self can become strained because being in relation with another is difficult as their capacities shift (2017, p. 211).

One way that a unique personal identity is maintained is through the co-authoring of narratives about oneself together with others in public and private spaces (Baylis 2017, p. 215). I know who I am in part because of the stories I tell and have been told, some of which began before I could tell my own stories. For example, when I decided to study philosophy, my parents helped me to tell the story of myself-as-philosopher by reminding me that when I was five years old I would often ask them, "Why are we here?" and that I decided it was to take care of the animals. These kinds of narratives help to construct me, and for me to understand myself as a unique person with particular characteristics that persist through time. My unique identity is thus co-authored (Baylis 2017, p. 216). This process is relational and, as Baylis describes, there are some constraints. I construct stories about who I am but I am not free to construct them in any manner I please. The stories are constrained by how others understand me. Others also construct stories about who I am as well, but these too are constrained by how I understand myself. Baylis calls this an "equilibrium constraint" where we try to find a balance in self-understanding and how others understand us.

Some forms of person-centred dementia care, such as reminiscence therapy are useful for countering the tendency toward fungibility by sharing biographical memories using photographs, memoires, or books that have been important to the person (Kontos 2005, p. 557). Reminiscence therapy can be useful, but as Pia Kontos notes, they are limited because they remain focused on recognizing the individual in so far as they are able to tell stories. At some point the person may no longer be able to tell their life stories, although others might be able to help them do so. Another aspect of 'me-ness' involves our embodied agency.

Pia Kontos considers the persistence of embodied agency as foundational to embodied selfhood. Although many philosophical accounts of what confers our non-fungible identity or 'me-ness' are based on cognitive capacities, Kontos notes that our embodied selfhood is perhaps the most enduring—both preceding and persisting after the development of complex cognitive abilities. Kontos follows Merleau-Ponty (2012) in the recognition of "nonrepresentational intentionality." Merleau-Ponty argued that the body itself demonstrates intentional agency in the way it is directed toward the world (cited in Kontos 2005, p. 560). For example, when bitten by a mosquito we don't have to stop and think about where on our body we were bitten and then decide to scratch. Instead we scratch the spot prereflectively, and the scratching manifests the agential intention to scratch. This kind of agency doesn't require learning; instead, it manifests the knowledge our body possesses (Kontos 2005, p. 560).

Kontos suggests that these "primordial capacities" are important but do not exhaust embodied selfhood. Especially with respect to those who have already lived a life the concept of habitus is also significant (2005, p. 562). Pierre Bourdieu describes habitus as the way we learn to embody culturally acquired behavioural propensities (1977, p. 72; 1990 p. 53). These propensities do not originate in the body, instead they are conditioned differently for members of different social groups (1977, p. 78; 1990, p. 54). Habitus is similar to embodied agency described by Merleau-Ponty because it occurs below the threshold of conscious awareness and is prereflective. It differs because it is learned rather than originating in the body. Pia Kontos suggests that the body is a fundamental source of selfhood that gives substance to the human self (2005, p. 567). Kontos suggests that prereflective embodied agency is particularly significant in those persons who experience severe cognitive impairment. For example, people with advanced dementia are often very sensitive to body language. Kontos says that patients who are experiencing advanced forms of dementia nevertheless retain their embodied selfhood. Their actions continue to demonstrate a prereflective intentionality toward the world. The habitual movements continue to express their self and the way that self developed a habitus in their social world. Because embodied agency is one of the most fundamental and enduring aspects of ourselves it might be particularly troubling when a person with dementia is treated as inert. To treat something as inert treats it as though it lacks agency or activity (Nussbaum 1995, p. 257). Olive, in *Case D*, was treated as inert when she was simply placed in front of the TV or a window and left there all day.

Agency is sometimes inappropriately identified as merely involving cognitive capacities, which turns agency into something akin to autonomy. If this occurs then people with dementia might inappropriately be treated as though they were inert and lack agency. Tom Kitwood contrasts the difference in the way a child's embodied agency and a person with dementia's agency are sometimes treated. In the case of infants, parents typically interpret their gestures as meaningful and respond to the gesture. This affirms that the infant's gesture "counts for something" (Kitwood 1993, p. 61) and enhances agency. In contrast, people with a diagnosis of dementia might be more likely to have their gestures met with no response. Kitwood suggests that such experiences might hasten the progression of the illness until a point where the individual appears to be vegetating (1993, p. 61). For example, the audit of Ontario long-term care facilities for people with dementia found that some residents would spend the whole day "parked" in chairs like "non-people" while the staff talked over them as though they were not there (Welsh 2018, Chapter 2).

In contrast, more respectful forms of therapy such as sensory therapy assumes that all behaviour has meaning. Gestures are treated as a basic substructure for articulating thoughts. These forms of therapy are an improvement and people with dementia can "come alive" after seeming to be inert. Kontos, however, suggests that they are limited by the focus on the articulation of cognitive elements such as thoughts (2005, p. 558). Kontos suggests that the body is not merely a substratum for thought. Instead, the body itself is an active, communicative, agent. According to Kontos a shift toward a greater recognition of embodied selfhood does not recommend new interventions. Instead it is a theoretical move that would shift the fundamental basis of caregiving. My suggestion here is that this theoretical shift would entail a shift in our conception of respect to the one I have described throughout this book. Even people with advanced dementia continue to "project their bodies with coherence" according to Kontos (2005, p. 565). Greater attention to embodied selfhood would recognize a broader range of behaviour as meaningful and can facilitate interpersonal relationships through cognitive decline. But wouldn't the kind of recognition of embodied selfhood Kontos describes entail reducing the person to their body? I do not believe that it does, though this claim requires a more complex account of what it means to reduce someone, or identify that person with their body.

When Rae Langton discusses reduction to body parts she is concerned with sexual contexts in which women are reduced to, and identified with, their bodies or its parts (2005, pp. 246–247). In Chapter 2, I described how in non-sexual contexts this reduction still happens. Particularly, an inherent part of medical research and clinical care involves a focus on the object body, its parts and the physiological functions thereof. This is not a moral failing as medicine would not be possible without the attempt to repair the body and its functions. If one attends *only* or *primarily* to the "malfunction" or disease category then this would likely involve a reduction to the body or its parts unless it was balanced by a humanizing context of moral respect along the other dimensions.

In the case of *reducing* a person to their body or its parts, the body is treated as inert rather than as an agent. Kontos' discussion differs significantly from the reduction to body parts because her idea of embodied selfhood centrally depends on embodied agency. The focus on the body's prereflective agency does not reduce the body to a mere malfunctioning, inert object to be managed, treated or cured from what Strawson calls the objective stance (1974). Instead, the recognition of the body as an agent engages with the intentionality of bodily movements and considers these movements to be meaningful. Continuing to recognize a person's embodied agency may mean that caregiving takes more time. A person with dementia often takes longer to perform tasks that a caregiver could quickly complete. The extra time is worth it to ensure the person is not disempowered or robbed of their own agency (Kitwood 1990, p. 182). As Kitwood argues, doing things for people with dementia can lead to de-skilling and a loss of confidence. Kitwood calls these processes part of a "malignant social psychology" that diminishes personhood and contributes to the development of dementia in a dialectical process.

The final forms of objectification that I consider in the context of caring for people with dementia are the denial or violation of subjectivity and silencing. People with cognitive capacities that differ from the normative ideal envisioned to apply to "healthy adults" often retain subjectivity and the capacity to communicate. They might, however, experience the world or communicate their experiences in ways that are unfamiliar or outside of social normative expectations. This might give rise to temptations to dismiss the emotions as "inappropriate" or expressive communication as "meaningless" or "just the disease talking." The description of second-person care-respect suggests that this would be a mistake. I end this section with a brief consideration of understanding across difference, or what Iris Marion Young calls "asymmetrical reciprocity."

A person has their subjectivity *denied* when they are treated as though they have no "inner life," experiences or emotions. A person's subjectivity is *violated* when the presence of emotions and experiences are recognized but treated as though they do not matter (Nussbaum 1995, p. 257). A person is silenced when they cannot speak, they are not allowed to speak, or they are not listened to (Langton 2005, p. 247). Sometimes the experiences of people with dementia are treated as insignificant because the person will not remember the experience for very long. The temptation to dismiss or violate their subjectivity might be particularly strong when task-oriented models of care are combined with too few staff and pressures to protect the bottom line. For example, the auditor of the long-term care facilities in Ontario observed a caregiver "shovelling food into a resident's mouth" because they needed to quickly free up the table to feed the next resident (Welsh 2018, Chapter 2). Additionally, the kind of listening involved in caring for those whose subjective experience differs from the normative ideal of healthy adults can be challenging. Utterances might come from a place that is unfamiliar to the caregiver. For example, Moira Welsh describes a training session for caregivers in which an elderly person with dementia believed she was ten years old and her mother was still alive. The caregivers knew this was not true and wanted to correct her mistakes of fact (Welsh 2018, Chapter 3). This might be the truthful way health care professional would interact with a healthy adult, but to do so in this case would fail to attend to the woman's experience and the shock that would likely ensue. Further, as David Sheard suggested in the training, reinforcing facts might fail to engage the meaning of the utterance.

Instead of asserting a fact in search of truth, the elderly woman might instead seek comfort, love or understanding (quoted in Welsh 2018, Chapter 3). In this instance, a fact-based response would not engage the woman's meaning even if it responds to the surface of the words.

Kitwood details several ways people with dementia are silenced. Sometimes health care professionals might talk over the person to one another as though the person with dementia was not even in the room (1990, p. 183; 2011, p. 93). During medical examinations a doctor might direct all of the questions to a family member without ever addressing the person with dementia themselves (1990, p. 184).[3] Kitwood suggests that these depersonalizing attitudes can contribute to the cognitive decline associated with dementia because they contribute to an invalidating and malignant dialogical process.

There are a number of forms objectification can take in the context of caring for patients with dementia. Metaphors that are used to describe Alzheimer's disease and pressures placed on health care professionals to provide "efficient" care might create a pull toward an objectifying perspective where a patient is reduced to a "slab of meat" to be "measured, pushed around, manipulated, drained, filled," as Kitwood describes (1990, p. 184). The role of moral respect in health care settings is to recognize the dignity of the individual. Recognizing a person's dignity involves attending to their equal moral value and their unique, non-fungible subjectivity and expressive agency.

When people are experiencing cognitive differences from normative ideals for healthy adults, respectful second-person engagements can be made more difficult as their emotional reactions, expressive utterances and agency may take unexpected and non-literal forms. It is possible to engage with people from a second-person perspective and to affirm their moral value across differences. For example, a nurse working in a long-term care facility described taking a particular man with dementia on walks in nature. This man loved watching the birds fly and he

---

[3]Havi Carel and Gita Györffy (2014) document how a similar form of silencing can take place in the care of children. The concerns I raise in this discussion are quite general and apply to many patients who are considered to lack the capacity to make autonomous decisions.

got particular joy from the sun on his face. The nurse was frustrated because her co-workers saw the walks as a waste of time because the man would not remember the walk in a few hours. The nurse was trying to articulate the importance of the walks and their effects on the man's well-being. She said, "It is true he might not remember the walk in an hour, but I don't remember what I ate for dinner a month ago. That doesn't mean my meal was unimportant or that I should not have taken the time to sit down with my family." In her wise words the nurse recognized that the man's subjective experience was itself valuable even if he might not recall the experience in the future. The kind of second-person respect I describe in this book affirms the equal moral value of each person and does not seek a reductive basis for this value in some narrow set of cognitive capacities. Nor does it reduce moral interactions to the exchange of moral reasons. It does require a communicative engagement that attends to the responses of the persons involved in the interaction, but it allows a broad range of forms of communication including body language and bodily agency.

The kind of respect I have described is characterized by what Young (1997) has called "asymmetrical reciprocity." Some accounts of moral respect assert that there needs to be a symmetry, or reversibility of positions, in order for people to respect one another. For example, Darwall's account of moral respect presumes the symmetry of equal moral status and authority. Darwall's description of empathy's connection to respect, which I look at next chapter, assumes further symmetry in that it involves a reversibility of positions. According to Darwall, empathizing with another involves an imaginative projection into the perspective of the other. Empathy, he says, involves feeling with the other and "seconding" their emotion as an appropriate response (1998, p. 269). Young argues against symmetrical views of moral respect because they collapse the differences among perspectives. Instead, she suggests respect requires careful listening and attention rather than imaginative projection (1997, p. 344). In order to engage with another person as a unique non-fungible locus of particular experiences one needs to recognize precisely that our positions are not symmetrically reversible in this way. Viola Cordova writes, "The concept of the singularity of the human species prohibits seeing the *other* as truly *an other*" (2007, p. 165; italics in original). This approach collapses the uniqueness of individuals, or the differences among social groups into a presumed singularity.

Although moral respect requires a reciprocal second-person engagement, this need not involve symmetrical reciprocity. Young argues that differences among people matter. We are not "so fully other that there are no similarities and overlaps," but there is a danger that projecting a view onto the other arrogates their view and collapses it into my own (1997, pp. 346–347). In place of symmetrical reciprocity Young suggests we think of moral respect as an asymmetrical exchange similar to gift-giving. In exchanging gifts we each offer something of value to the other and we open up a reciprocal relationship. The gifts exchanged cannot be identical, however, or else the practice would not be one of gift-giving; it might instead be a purchase (1997, p. 343). The account of moral respect that I describe provides further support to Young's argument. If moral respect involves acknowledging the dignity of a person and dignity described the non-fungible, inimitable subjectivity of that particular person, then the idea of symmetrical reciprocity is further called into question.

Rather than an imaginative projection, Young suggests a moral humility that assumes we cannot "see things from the other's perspective" (1997, p. 350) although we can understand the other by careful listening. Young's discussion is useful, although she sometimes seems to presume verbal exchanges with the use of "listening" to the other. The account I have described in this section would extend beyond listening to the surface meaning of verbal exchanges and include a variety of responses including body language, emotional expression, embodied agency and so forth. The kind of moral respect I advocate for in this book takes time and patience in interactions. One of its primary drawbacks is that it is not easily reducible to a check list or a set of standardized tasks that can be completed as quickly as possible and submitted to managers for monitoring. The attempt to do so would transform alleged "respect" into a third-person concern, which would mean it would not be moral respect at all according to the view I have described. I think this is a happy result, as it reflects Kant's original description of the difference between "dignity" and "price." While it is important to think about saving resources and making sure there is enough to go around; for some kinds of things, like human dignity, it is disrespectful to trade them off against economic concerns.

3.5   Conclusion

In the last chapter I argued that dignity does not need to be grounded in some psychological capacity or set of capacities. Instead, dignity describes a type of value. In this chapter I have tried to fill in an understanding of dignity and respect for non-autonomous patients. We have dignity because we are unique and non-fungible beings with inimitable subjective and emotive experiences, relationships, and embodiments. We have particular non-interchangeable subjective perspectives. We grow up in particular relationships to our families, friends, and societies; we have particular life stories that differ from others, even as they share similarities in some ways. We have embodied agency and our bodies are directed toward the world. All of these aspects contribute to our "me-ness" that makes us who we are. In addition to these aspects, some of us will become, are or once were autonomous, while others of us will never be autonomous. For those of us who are autonomous, our self-determination is often a central element of what we take to be important about ourselves. Losses or changes along a single dimension of what makes us a unique and irreplaceable individual are not sufficient to cause a loss of self or a loss of dignity. On the account I have offered, those who are not capable of autonomous self-direction retain dignity and are owed moral respect in affirmation of their equal moral value. In the next chapter I consider whether this account would also apply to autonomous individuals.

References

Baylis, Françoise. 2017. Still Gloria: Personal identity and dementia. *International Journal of Feminist Approaches to Bioethics* 10 (1): 210–224.

Bourdieu, Pierre. 1977. *Outline of a theory of practice*, trans. Richard Nice. Cambridge, UK: Cambridge University Press.

Bourdieu, Pierre. 1990. *The logic of practice*, trans. Richard Nice. Stanford, CA: Stanford University Press.

Carel, Havi, and Gita Györffy. 2014. Seen but not heard: Children and epistemic injustice. *The Lancet* 384 (9950): 1256–1257.

Cordova, Viola F. 2007. *How it is: The Native American philosophy of V. F. Cordova*, ed. Kathleen Dean Moore, Kurt Peters, Ted Jojola, and Amber Lacy. Tucson: University of Arizona Press.

Darwall, Stephen. 1998. Empathy, sympathy, care. *Philosophical Studies* 89 (2–3): 261–282.

Darwall, Stephen. 2006. *The second-person standpoint*. Cambridge, MA: Harvard University Press.

Dillon, Robin. 1992a. Care and respect. In *Explorations in feminist ethics: Theory and practice*, ed. Eve Browning Cole and Susan Coultrap-McQuin, 69–81. Bloomington and Indianapolis: Indiana University Press.

Dillon, Robin. 1992b. Respect and care: Toward moral integration. *Canadian Journal of Philosophy* 22 (1): 105–132.

Dillon, Robin. 1997. Self-respect: Moral, emotional, political. *Ethics* 107 (2): 226–249.

Kitwood, Tom. 1990. The dialectic of dementia: With particular reference to Alzheimer's disease. *Aging and Society* 10 (2): 177–196.

Kitwood, Tom. 1993. Toward a theory of dementia care: The interpersonal process. *Aging and Society* 13 (1): 51–67.

Kitwood, Tom. 2011. Dementia reconsidered: The person comes first. In *Adult lives: A life course perspective*, ed. Jeanne Katz et al., 89–99. Bristol: Policy Press.

Kontos, Pia. 2005. Embodied selfhood in Alzheimer's disease. *Dementia* 4 (4): 553–570.

Langton, Rae. 2005. Feminism in philosophy. In *The Oxford handbook of contemporary philosophy*, ed. Frank Jackson and Michael Smith, 231–257. Oxford: Oxford University Press.

Lindemann, Hilde. 2002. What child is this? *The Hastings Center Report* 32 (6): 29–38.

Merleau-Ponty, Maurice. 2012. *Phenomenology of perception*, trans. Donald Landes. New York: Routledge.

Noddings, Nel. 1984. *Caring: A feminine approach to ethics and moral education*. Berkeley: University of California Press.

Nussbaum, Martha. 1995. Objectification. *Philosophy & Public Affairs* 24 (4): 249–291.

Ruddick, Sara. 1995. *Maternal thinking: Toward a politics of peace*. Boston: Beacon Press.

Schwartz, Meredith Celene. 2007. Growing concerns: Prenatal genetic risks and trust. In *Risk and trust: Including or excluding citizens?* ed. The Law Commission of Canada, 79–101. Black Point, NS: Fernwood Publishers.

Scully, Jackie Leach. 2008. *Disability bioethics: Moral bodies, moral difference.* New York: Rowman & Littlefield.

Shenk, David. 2000. Biocapitalism: What price the genetic revolution? In *Contemporary moral issues: Diversity and consensus*, ed. Lawrence Hinman. Upper Saddle River, NJ: Prentice Hall.

Spelman, Elizabeth. 1978. On treating persons as persons. *Ethics* 88 (2): 150–161.

Strawson, Peter. 1974. *Freedom and resentment, and other essays.* London: Methuen.

Tronto, Joan. 1993. *Moral boundaries: A political argument for an ethic of care.* New York: Routledge.

Welsh, Moira. 2018. The fix: One Peel nursing home took a gamble on fun, life and love. The most dangerous story we can tell is how simple it was to change. *The Toronto Star*, June 20. http://projects.thestar.com/dementia-program/. Accessed 21 June 2018.

Young, Iris Marion. 1997. Asymmetrical reciprocity: On moral respect, wonder, and enlarged thought. *Constellations* 3 (3): 340–363.

4

Respect and the Lived Experience of Illness

Abstract Schwartz describes phenomenological accounts of embodied agency and the objectification and alienation that are part of many experiences of illness. In health our bodies are the centre of our world but rarely the centre of our attention: there is a seamless unity between the object-body and the body as subject. Experiences of illness focus our attention on the object-body which is thematized as a problem and a limit to our agency. In disabilities present since birth this involves not a changed embodiment, but a lack of fit with the habitus of the social world. Phenomenological accounts of illness and disability explain the importance of moral respect in health care contexts from an autonomous patient's perspective. Respect does a better job of dealing with objectification than empathy because empathy inappropriately adds to the emotional labour of health care professionals whereas respect does not.

Keywords Phenomenology of illness · Empathy · Embodiment · Objectification · Alienation

© The Author(s) 2019   **87**
M. C. Schwartz, *Moral Respect, Objectification, and Health Care*,
https://doi.org/10.1007/978-3-030-02967-8_4

4.1   Introduction

The account so far leaves out autonomous patients. Would there be any advantage for autonomous patients if we considered moral respect to play the role of countering objectification? I believe there are advantages for autonomous patients in this account of the role played by moral respect. Autonomous patients would benefit from a more expansive account of respect that attends to a variety of forms of objectification because serious illnesses are often experienced as objectifying from a first-person perspective. When one feels alienated from one's own body as a result of illness this can be existentially unbearable. I draw on phenomenological accounts of illness to explain how an account of moral respect as countering objectification can include a broader consideration of experiences of illness; experiences that are of central importance to patients who are confronting new diagnoses, the onset of disability (or disabilities) and chronic medical conditions. Attending to respectful relations from the interactive stance can help autonomous patients return to a fuller sense of their own subjectivity and moral value. In this chapter I focus only on those patients who retain the capacity to make their own medical decisions to show that an expanded account of moral respect would be advantageous *even* to those who are covered by the well-known principle of respect for autonomy.

Medicine involves, at least in part, adopting an objective attitude to consider the physiological functioning of a patient's object-body. The experience of illness itself and the encounter with complex health care systems can result in patients taking an objective attitude toward themselves. They might find that the nature of their illness focuses their attention on the object-nature of the body over the lived body. These experiences taken together can be deeply objectifying where the patient might come to feel as though they are being reduced to an object. We can adopt more than one perspective on ourselves and others, however. One of the reasons we need an account of respect in bioethics is to counter this objectification and help patients return to their subjectivity. Respecting another person, involves attending to their uniqueness and their inherent value (or dignity). To respect another is to recognize the other as more than a mere object among objects. Respecting another affirms that they are the locus of perspectives, values, affects, and experiences that are inimitably theirs. Respect is important in bioethics so that the experience of illness becomes more existentially bearable for the patient. Making the experience of illness tolerable, perhaps even livable, is essential to healing the patient and not merely curing the body (Toombs 1995, p. 20). I argue that respect is particularly important in medical contexts from a patient's perspective in part because the experience of illness has an inherently objectifying aspect from the first-person perspective. Many phenomenological accounts of the experience of disability and illness describe the way in which patients become focused on the object-body in a way that disrupts or alters the experience of the body as lived. My claim here is that the experience of illness itself involve an inherent element of objectification of a patient. My focus in this chapter will be on physical illnesses rather than mental illnesses because my focus is on changes to the object-body. It is possible that some of these descriptions would apply to people with mental illnesses as well. Although the phenomenological changes involved in mental illnesses differ, the experience of being objectified during encounters with medical bureaucracies will likely have similarities.[1]

In Sect. 4.4, I consider how encounters with medical treatment can increase a patient's sense that they are a mere number or "slab of meat" and I argue that patient experiences of illness within complex medical bureaucracies make respect into a particularly important moral concept in health care institutions. Because these experiences involve various forms of objectification, I suggest that respect ought to involve a variety of responses. Recently it has become fashionable to describe empathy as important for attending to patient experiences. In the fifth section I describe some reasons that I believe a full account of respect might be preferable to suggestions that health care professionals ought to empathize with patients. In particular, I claim that this suggestion could be detrimental from the perspective of the health care professional's

---

[1]The objectification involved in encounters with health care professionals might be even greater in cases of mental illness due to stigmatization.

experience and well-being. One advantage of the account of respect that I offer is that it can better engage with a patient's experience of the bodily and subjective changes involved in illness but without requiring a health care professional to share in these experiences.

One of the claims I wish to make in this chapter is that objectification occurs in illness not merely from the outside but also from within as a result of illness itself which can involve self-objectification and alienation from one's own body. In this chapter, I add to the account developed in Chapters 2 and 3 a consideration of respect in health care as it relates to the first-person perspective of the person experiencing illness.

4.2   Embodied Selves

In this section and the next I consider changes to one's embodied agency that occur in experiences of illness. This section focuses on embodied agency when people are healthy and the next section focuses on some changes to embodied agency that occur in illness.[2] The distinction between disease and illness mirrors the distinction between the third- and first-person stance I have relied on throughout this book. Medicine provides us with an objective third-person account of disease. In contrast, phenomenology privileges first-person experiences of illness (Carel 2013a, p. 10). While medicine focuses on the physiological aspects of disease or disorder phenomenology considers the experience of illness of what it feels like to have a particular malady (Carel 2013a, p. 13). In Chapter 2 I described how medicine has an inherent pull toward the objective, third-person stance. This results from the naturalized perspective medicine adopts with respect to illness. A naturalized approach to illness views disease as the result of natural or physical facts alone (Carel 2013a, pp. 9–10).

---

[2] I am indebted to Ley (2017) for pointing me toward the relevance of embodied agency and the distinction between disease and illness.

Some might object that medical care does attend to the unique, subjectivity and particularity of the patient; after all, doctors must rely on patient reports about their symptoms, pain and how they feel in response to the treatment. But, as Richard Baron notes, there has been a marked shift away from making a patient's symptoms central to clinical nosology. Baron suggests that in the late eighteenth and early nineteenth century, diseases were classified by the experiences of symptoms as they were reported by patients. But by the mid-nineteenth century medical nosology began to take diseases to be an "anatomicopathologic fact" (1985, p. 606). Baron writes, "We tend to see illness as an objective entity that is located somewhere anatomically or that perturbs a defined physiologic process. In a profound sense, we say that such an entity 'is' the disease, thus taking illness from the universe of experience and moving it to a location in the physical world" (1985, p. 606). Further, as Kay Toombs notes, the perspective medicine takes on the body is mainly a "Cartesian" paradigm that involves a dualistic split between the mind and the body. The body is then conceptualized in mechanistic terms (Toombs 1988, p. 201). This approach within medical epistemology has been successful and with it has come advances in new interventions to repair the body and its parts. Although Baron agrees that this perspective on medicine has helped bring about medical advances, nevertheless, the classification of the disease by the anatomic or physiological pathologic fact makes physicians focus on the disease object instead of the sick person (1985, p. 607).

Because of the successes of modern medicine phenomenologists interested in experiences of illness do not typically want to replace the naturalistic approach within medicine. Instead, they are interested in making explicit which aspects of illness are occluded by the naturalistic approach. The goal is to augment medical understandings rather than replace them (Carel 2013a, p. 10). Phenomenological accounts re-centre the first-person experience of patients who are living through the illness. This supplementation is important because the mechanistic, third-person, objective descriptions of bodily dysfunction do not capture the experience of illness as lived. As Toombs notes, she does not experience the lesions in her brain or abnormal reflexes. Instead, she writes, "my illness is the impossibility of taking a walk around the block or carrying a cup of coffee from the kitchen to the den" (1995, p. 10). We experience illness as changes in our lives not as changes in our physiology.

Phenomenological accounts begin with the notion of the lived body developed by Maurice Merleau-Ponty (2012). Descriptions of the lived body take embodiment as our being-in-the-world. I always find myself embodied within the world. As an embodied subject I don't experience my body as an object *of* the world. Instead, the body as I live it represents my perspective *on* the world (Merleau-Ponty 2012, p. 94). My body differs from other objects in the world because it is always with me and is always perceived from the same angle. Merleau-Ponty writes that my body's permanence "is not a permanence in the world, but a permanence on my side" (2012, p. 93). I observe external objects with my body by moving them around to inspect them. But I cannot spread my body in front of me to examine it from multiple perspectives (2012, p. 93). We experience the world through our bodies, but we don't observe our bodies as an object of experience. Although we can see our eyes in the mirror this differs from the sensation of seeing itself (Merleau-Ponty 2012, p. 94). As Toombs notes, it is impossible to perceive the sensation of seeing. She writes, "all that I can grasp is my 'seeing of' this or that object, not the actual sensory activity itself" (1988, p. 215). The body is always present as the centre and reference of my world but it is "the inapprehensible given" of first-person experience (Toombs 1988, p. 215). Havi Carel describes the healthy body as more than a mere thing among things. The body is the foundation of all experience, which makes perception possible, but the body itself typically remains peripheral to our awareness (Carel 2014, p. 24). Although all of our experiences occur through the body our attention is on the objects in the world rather than on the body and there is a "seamless unity between the body as object and the body as subject" (Carel 2014, p. 31). The experience of lived body is a first-person experience where we experience through the body but do not attend to the body itself.

The lived body is directed toward the world and exhibits bodily intentionality (Toombs 1988, p. 204). Our agency is embodied and we demonstrate this corporeal non-reflective intentionality in the movements of the body (Kontos 2005, p. 561). Bodily agency is non-reflective because we do not need to coordinate our actions by thinking about the body part and then forming an explicit decision to move that part. Instead, as Pia Kontos describes, we intend a certain outcome by our actions and these are "spontaneously distributed amongst the appropriate parts" of our bodies (2005, p. 561). Bodily intentionality arises spontaneously and, as Merleau-Ponty writes, "in the very first attempts at grasping, children do not look at their hand, but at the object" (2012, p. 150). The body is oriented toward the world and its objects as possibilities for actions. As Toombs writes, "movement presents itself to the body as a practical possibility, a sphere of action" (1988, p. 204). Embodied agency is directed at the world in a seamless manner we typically need not reflect upon consciously. Instead, bodily movements are spontaneously coordinated toward some goal.

When certain activities have become habitual for us, our body is usually transparent as we engage in the activity. As I am learning a new activity (for example how to swim) I might attend to my body as an object trying to learn a new activity. Once I have been swimming for several years, however, my body fades to the background. I don't attend to the body that is swimming but instead attend to the swimming itself (Toombs 1988, p. 216). I have a goal to get to the other end of the pool, perhaps within a certain time. There is an 'intentional arc' between me and the pool where my body orients to help me reach my goal and there is a seamlessness between my intention to reach the other end of the pool and the movements of my body that effect this goal. If I focus on my body *as* a body trying to swim then this disrupts the typical ease with which I glide through the water. I might find myself making wasted movements and sputtering about in the water. The healthy, lived body is the locus of our activity but typically it is not an object of our attention.

In health, Carel suggests, we have a "taken-for-granted unreflective, disinterested, tacit belief that we will accomplish the things we have done in the past" (2013b, p. 182). This gives rise to a bodily confidence we rarely reflect upon. Bodily confidence forms part of the background to our experiences. This background certainty isn't completely transparent because the world resists us. Even healthy people will find that it is difficult to open a new jar or lift a heavy object. If we have performed these activities in the past, we develop a confidence that our body will be able to complete them again in the future.

In addition to our agency originating in bodily intentionality and the bodily confidence that results, our embodiment also inscribes social understandings, as I described in Chapter 3. Pierre Bourdieu (1977, 1990) considered how these social understandings are embodied through habitus. This aspect of embodiment is also pre-reflective and takes on a form of "feeling of right behaviour" rather than explicit knowledge that can be consciously articulated (Scully 2008, p. 65). Bourdieu's concept of 'habitus' refers to "patterns of being and doing" that are acquired within particular social fields. These social fields have tacit rules that give rise to habitual patterns of behaviour that exhibit a pre-reflective understanding of the norms. As Jackie Leach Scully describes, "Habitus is a pretheoretical, prereflexive knowledge that we absorb from the behaviour and practices that are demonstrated, rarely articulated, by the people around us" (2008, p. 65). The structures of the habitus are inscribed early on and are powerful because they are outside of our conscious control. We develop a feeling of "right behaviour" but we might not be able to articulate it in words. These prereflective background meanings nevertheless shape our conscious reasoning. According to Scully, habitus matters for moral agency because practices that conform to the habitus will seem "fitting" and "morally good" (Scully 2008, p. 66). Part of the social rules we come to embody through the habitus will include moral ideas and emotions. Although agents might offer propositional justifications for their actions, the reasons offered "feel right" because they accord with the habitus (Scully 2008, p. 67). Our agency is embodied in ways that originate both from within the body itself and from our social fields. As I described in Chapter 3, our embodied agency will include experiences we have that either develop a robust sense of self-respect or damage our basal self-respect. Social values are inscribed in our embodied agency.

To sum up, phenomenological accounts make us aware of our embodied agency. The body is the centre of our world but rarely the focus of our attention. Once we have learned a new skill we don't have to concentrate on the bodily performance of an action and we become confident that we will continue to exercise those skills. Habitus structures social norms into the movements of the body. Although these culturally acquired patterns of behaviour are learned and do not originate in the body, the learning takes place at a prereflective level and is difficult to explicitly articulate. Nevertheless, our conscious and explicit reasoning is conditioned by the prereflective background meaning of habitus.

4.3   Lived Experiences of Illness and Objectification

One element of phenomenological accounts of experiences of illness involves a changed relationship to the body. The changes to the body and bodily functioning through illness affect our way of being in the world and the meaning of our experience. As Kay Toombs explains, "Illness is experienced not so much as a specific breakdown in the mechanical functioning of the biological body but more fundamentally, as a disintegration of his 'world.' This is not surprising when one recognizes that illness-as-lived is a disruption of lived body" (1988, p. 207). For example, the experience of a migraine headache is not simply having a pain in one's head. It also involves difficulty concentrating on what one is reading (Toombs 1988, pp. 207–208). Although there are several ways these changes can manifest, I focus on the changes as they relate to objectification.[3] Illness can make the body, that was once taken for granted, come to the forefront of our attention *as* an object of contemplation that is thematized as a problem. Illness disrupts the seamless unity between the object-body and lived body in ways that often focus our attention on the former. The precise character of this

---

[3]Of course, phenomenological accounts of illness (as a general phenomenon) are much more nuanced and complex than the narrow focus on objectification presented here. For example S. Kay Toombs (1988, 1995) describes a rich account of changes to lived spatiality and temporality as well as the experience of the uncanny in illness. Havi Carel argues that the experience of illness changes our sense of bodily certainty, can involve a loss of continuity (2013b) changes in embodiment, the meaning of objects in our environment and our being in the world (Carel 2014, p. 22). These accounts of illness as a general phenomenon are enriched further by thick, nuanced accounts of particular illnesses, mental illnesses, and disabilities.

change is variable and depends on the nature of the illness: its onset, duration and intensity. For example, the changes to embodiment for an illness that has a short duration but is intensely debilitating will have a different character from a chronic illness in which the changes in one's ability degrade over an extended period of time. Different forms of illness provide us with differing opportunities to anticipate or adjust to bodily changes. The differences associated with different types of illness might mean that some of these experiences are absent for certain conditions, so we should not presume that all bodily differences will be accompanied by increased perception of the object-body. For example, if an illness or disability has been present from birth, one might experience this difference in embodiment as transparent since this is how one has always interacted with the world. My intention in this section is not to provide a complete phenomenological account of illness; instead, I draw on only some select parts from such accounts provided by others to highlight some of the ways in which experiences of illness often involve objectifying elements and how this can lead to deeper self-objectification that can be, alienating and perhaps at the extreme "existentially unbearable" as Carel writes (2014, p. 32). Moral respect in health care contexts helps patients to return to their fuller selves.

Illness often destabilizes the structure of experience and the nature of our being in the world. In illness, the biological body interferes with the 'intentional arc' of the person. The body that has typically been the means for reaching our goals might now come to seem like an obstacle to those goals. Carel and Toombs describe this process as a process of objectification in illness. The negative and limiting aspects of illness cause our attention to be focused back toward our own body as an object: it might come to seem as a thing among things, which is a form of self-objectification. Rather than seamlessly functioning to accomplish tasks and goals, the body in illness can thwart our plans and make some actions impossible. "Illness disrupts the fundamental unity between body and self which characterizes the lived body," according to Toombs (1988, pp. 214–215). Instead of being an instrument for one's actions the biological body is perceived as an object, as physical, and perhaps itself as an obstacle.

In illness "the body can no longer be taken for granted or ignored. It must be explicitly attended to in various ways," a process Toombs describes as objectification (1988, p. 214). Rather than being the seamless medium of our agency that is transparent to our awareness our body comes to the fore as an object that itself might require attention, tending and care. If our illness requires testing and imaging of our physiological processes or organs then these aspects of our object body, once literally invisible, become the centre of our attention in ways we might not have imagined before (Carel 2013b, p. 191).[4] Moreover, the shift from lived body to object body does not occur in a neutral context. Instead, as Carel notes, in illness "the body becomes *explicitly thematized* as a problem" (Carel 2013b, p. 191; italics in original). Our tacit taken for granted attitude toward our body—as we expect it to perform habitual movements, to be pain-free or to allow us to concentrate—is disrupted. The tacit attitude is "replaced by an explicit attitude of concern, anxiety, and fear" (Carel 2013b, p. 191). This can lead to a negative attitude toward our own body in illness. In some cases this can lead to a sense of alienation between self and body. Alienation from the body occurs when we experience the body as other-than-me (Toombs 1988, p. 216). This disruption is a disturbance in our embodied consciousness and world that "strikes at the very heart of the 'I am'" (Toombs 1988, p. 207). When we experience an alienation between self and body we experience the body as an object, something that is "outside of my subjectivity in the midst of a world which is not mine" (Toombs 1988, p. 216).

Havi Carel describes how these changes in one's embodied experience of self can lead to a sense of bodily doubt (Carel 2013b, pp. 189–190). Bodily doubt occurs when one can no longer take the body's capacities for granted. Whereas bodily certainty involves a tacit belief we can accomplish the things we have done in the past, bodily doubt focuses

---

[4]Some of the changes to embodied agency described by Toombs and Carel also occur in aging. I do not think aging is an illness, although the processes of aging are often medicalized. I think that Toombs and Carel describe phenomenological experiences of illness but I do not think the claim is that these experiences only occur in illness. For example, sexual objectification of women might also involve a greater attention to the body, inhibited intentionality and discontinuous unity (Young 2005, p. 35). I am indebted to Susan Sherwin for drawing my attention to the connection to aging.

on present incapacities that are cut off from future goals (Carel 2013b, p. 189). This shift means that living bodies become "fragile physical entities (inert, self-facing)" that are dislocated from embodied agency (Carel 2013b, p. 189). As Carel writes, "The focus shifts from experiencing oneself primarily as an intentional subject to experiencing oneself as a material object" (2013b, p. 190). In bodily doubt the activities that were once taken for granted become achievements. These changes have significant effects on our being in the world.

As Toombs explains, the experience of illness is first and foremost one of dis-ability, "the inability to engage the world in a habitual way" (1988, p. 207). Even relatively minor illnesses have this character. When we have the flu, for example, we might find the habitually taken for granted activity of getting out of bed is now difficult and laborious. The last time I had strep throat I watched a film in which a professor was enthusiastically presenting to the class. Even though I had done that myself only a few days earlier, in the midst of illness it seemed unimaginable to me. I could scarcely believe that I might be doing the same thing again in a few weeks. Although even minor illnesses involve an experience of self-objectification, they are temporary and so are less likely to involve a thoroughgoing shift of being-in-the world that will lead to alienation.

In the case of enduring differences such as impairments that have been present since birth, objectified embodiment might not involve changes per se. In such cases attention to the object body might result from a lack of fit with one's social world. Scully describes how most people born with impairments grow up in settings where their body is not normative and they cannot "effortlessly" slip into the social fields they experience (2008, pp. 68–69). Habitus instils not only a sense of "fitting" behaviour it also inscribes an equally prereflective and powerful sense of "*illegitimate* modes of bodily representation" (Scully 2008, p. 69; italics in original). This does not mean that people with impairments cannot play the game at all. When people say things like, "But I never think of you as disabled" they are pointing to success at conforming to normative expectations. Scully maintains that for people with disabilities what they cannot do is inhabit these normative expectations with the same degree of unthinking ease (2008, p. 69).

Scully gives the example of children born with limb abnormalities as a result of thalidomide. As these children learned to move they would do so using their limbs in unusual but functional ways. Other people were often uncomfortable seeing the children use the "wrong" limb or walking on their hands rather than on their feet (Scully 2008, p. 70). Doctors offered prosthetics which frustrated the child's agency but made others more comfortable with their appearance. The children resisted the changes because the prosthetics were awkward and made their movements difficult. Eventually the doctors stopped asking the children to use the prosthetics. In cases such as these it is not so much bodily *changes* that are objectifying. When a disability has been present since birth it might be the reaction of others that frustrates one's seamless agency. Scully suggests that the discomfort that other people feel at non-normative ways of being in the world are "enormously strong" and gain their power from "the prevailing habitus and its internalized rules" (2008, p. 70). In such cases what might be required are changes to the tacit norms of the social fields themselves.

In sum, the experience of illness, whether acute, progressive, or chronic, can involve self-objectification by the ill-person. The extent to which these experiences will call for an interactive second-person response of respect will thus depend on the severity and duration of the illness. To return to the idea of balance introduce in Chapter 2, an illness that is particularly acute or an illness that is chronic will often lead to greater and more pervasive forms of objectification. Hence an engagement with a patient in ways that counter these forms of objectification will be more important. Carel explains that illness causes a rift between the lived body and the biological body. In health, the biological and lived body typically cohere. The experience of the lived body is a first-person experience where we experience through the body but do not attend to the body itself. Although all of our experiences occur through the body, our attention is on the objects in the world rather than on the body and there is a "seamless unity between the body as object and the body as subject" (Carel 2014, p. 31). But in illness the biological body interferes with the "intentional arc" of the person and the body that typically has helped us reach our goals might now come to seem like an obstacle. This process is a process of objectification of the body in illness, as Carel describes it. The negative, unwanted and limiting aspects of illness cause our attention to be focused back toward our own body as an object: it might come to seem like a thing among things, which is a form of self-objectification.

Changes in our way of inhabiting our bodies in illness are frequently accompanied by changes in our agency and autonomy. Carel notes that in illness, "[the] bodily foundations of autonomous adulthood are often removed, revealing the tentative and temporary nature of these foundations" (Carel 2014, p. 24). These changes might partly explain why health care ethics has traditionally focused on respect for autonomy understood as medical decision-making: if our bodies feel out of our control it might be comforting to have our decisions accepted as it provides us with a means of controlling at least some aspects of an out-of-control situation. Thus I would not ignore the importance of having one's medical decisions accepted. I do not think that accepting treatment decisions is sufficient for helping patients who are experiencing changed embodiments, and hence changed worlds. With a fuller phenomenologically grounded account of the importance of moral respect we can see shifts that occur in patient agency in everyday contexts beyond clinical decision-making. To help patients live well with their illness health care professionals ought to help patients adjust to these changes, and the account of respect that I have described is one important way to do so.

Since illness has profound effects on our embodiment, our agency, and bodily certainty and these changes can make us alienated from our physical body, which might come to seem outside of our subjectivity, it is little wonder that respect has been considered central to ethics in health care contexts. One of the important functions of moral respect is to recognize that we are more than a mere object among objects. The account of respect that I am proposing would suggest that health care professionals respect their patients when they help them live well with the embodied changes and regain a sense of first-person subjectivity. When the professional helps a patient develop new ways of exercising agency that repairs the rupture between lived body and object body the respectful relationships have helped counter the self-objectification and alienation involved in experiences of illness. Unfortunately, as I describe in the next section, encounters with health care professionals in complex medical bureaucracies too often has the effect of increasing alienating forms of objectification of a patient's physical body.

4.4   Objectification in Encounters with Others

The experience that the ill person has with friends, strangers, medical professionals and complex bureaucratic medical systems can contribute to the objectification of the body in illness. Carel describes the ways in which the medical emphasis on the body as an object can contribute to the rift between the lived body and the biological body for the ill person in ways that can alienate the patient from her body. For example, as the patient waits to hear the results of a blood test or to have the images produced by a CT scan interpreted for her, the typical relation of a subject to her body shifts. Typically, we think we have privileged first-person access to our own bodies. But the results of medical tests and imaging are objectifying because the person is in no better position to examine her own bodily states than is the objective third-person observer (i.e. the physician). The patient might require the physician's expertise in order to interpret the results of these tests that report on her bodily functioning. This creates a distance that might encourage the patient to see their body as an object (Carel 2014, p. 32).

During medical appointments a doctor's focus is on the object body and its functions. The boundaries of the body, which typically protect my inviolability, are penetrated by instruments or by imaging equipment that peers below my skin. Moreover, this focus is typically negative and test results measure how much the body's functions deviate from the normal range (Carel 2013b, p. 191). Since normal functioning is thought to indicate health patients will often become focused on even small fluctuations in these measurements. For example, when my friend Tara was undergoing chemotherapy I would sometimes accompany her to the treatments. As we got to know the other patients we would say hello and ask how they were. At first I was surprised that they would usually respond by telling us about the latest measurement for their tumour markers. Typically a friendly, "hi, how are you?" will be met with a subjective response such as, "I'm fine," or "I have a pain in my shoulder." For these patients the intensive focus on their object-body and the objective measurement of its function came to replace the subjective response.

Further, to cooperate with the physician, the patient has to describe and explicitly attend to the body as an object. Patients are expected to focus on an "objective" report on their bodily sensations, their success at adhering to a treatment regimen, and any changes they notice as a result of treatment (Leder cited in Toombs 1988, p. 217). Naomi Scheman points out, these kinds of reports about a patient's symptoms are considered something only the patient can know, something to which the patient has privileged access (2009, pp. 118–119). The information a patient provides is not attended to primarily as an expression of the subjective experience of the patient. Instead, it is treated more akin to raw data.[5] The physician will then work with the raw data, interpreting it through an expert scientific lens to come up with a diagnosis or decide how the report should be coded for it to be clinically useful. So a patient might experience the body not only as an object but as "reduced to a malfunctioning biological organism" (Toombs 1988, p. 217). Rather than countering the objectification involved in diagnosis, being asked to report on one's symptoms while having only some parts of that report taken as relevant information might be experienced as a further form of objectification, namely silencing. Although silencing is typically thought of as treating someone as though they cannot speak at all, a form of silencing arises in these exchanges as only certain kinds of speech are welcomed. If a patient would prefer to describe a more first-person account of the changes not to their body but to their lives this will often not be welcomed by health professionals who might consider this kind of speech to be a waste of time.

Finally, the manipulation of bodies in hospitals and clinical settings can contribute to the sense of objectification. As Carolyn Ells notes,

---

[5]An example of how nuanced and particular reports about the subjective experience of patients can come to be treated as raw data for medical decision-making is provided by Paul Brodwin's description of the staff meetings among psychiatrists and case managers working to provide Assertive Community Treatment (ACT) to people with mental illnesses (Brodwin 2013, p. 63).

bodies in health care settings are "prone to being exposed and grasped with little or no explanation or apology by a variety of strangers who implicitly make some professional *claim* on those bodies" (2001, p. 611). In my own experience with infertility it was the objectification in the medical encounter that led me to feel disrespected. I had previously interpreted my ability to avoid pregnancy as a sign of responsible agency. When I tried to become pregnant without success I had to reassess my previous sense of control over my body and reinterpret it in light of a diagnosis of infertility. The tests to determine the cause of infertility focused my attention on my womb and fallopian tubes, parts of my body I almost never experience. My experience of these parts was negative because the hysterosalpingogram (HSG) was painful in addition to focusing my attention on the biological dysfunction. All of these objectifying elements might have been tolerable if I had experienced more respect from the health professionals at the fertility clinic. The combination of the inherently objectifying tests and diagnosis was made worse by the experience in the clinic, however. The part that ultimately made me feel humiliated and disrespected was when I was left alone in a room of fainting couches, that had not been explained as there for my use. I was alone in nothing but a hospital gown, feeling faint and unable to walk. I was not sure whether I could use the couch and there was no one to ask. It was the experience of being exposed and abandoned that made me feel dehumanized and reduced to a mere biological object regarded as a "unit of work." Had one of the health care professionals simply stopped into see whether I was recovering I might have felt differently. As it was, I had to wait for an orderly doing his rounds to even be able to request a glass of water.

The objectifying perspective of medicine is powerful. Indeed, it is so powerful that many patients come to think of themselves in objectifying ways. Adopting an objectifying stance does not preclude taking other kinds of stances, however. We can attend to others in more than one way and we can attend to ourselves in more than one way as well. Indeed, Carel argues that although we can come to think of our bodies as objects when we experience illness, "[w]e cannot actually view ourselves objectively in any sustained sense..." (Carel 2014, 32) and the attempt to maintain this stance towards ourselves would be "existentially unbearable." Carel proposes that health care professionals, who often value objectivity, should be aware of the self-objectifying nature of illness. In the next section, I argue that the importance of respect within health care ethics is to remind health care professionals that they, like the patients, need to return to a perspective that affirms the patient's embodied subjectivity. In engaging with patients from a respectful interactive stance health care professionals can help ease the existential weight of facing a serious medical diagnosis. This can help patients regain a sense of first-person subjectivity rather than furthering the experience of illness as objectifying and potentially alienating.

4.5   Respecting Through Illness, Empathy and Sympathy

Subjectivity involves our perceptions of the world and our interpretations of those experiences. The irreducibility of our individuality is expressed in our subjectivity. It is for this reason that I claim that respect ought to be understood as attention to multiple forms of objectification: respect is the recognition of the inalienable, absolute, moral value of each unique individual. When we respect an other we see the other as non-fungible and irreplaceable. This is in contrast to one of the perspectives in medicine that considers the object-body as interchangeable with other bodies. Attending to the uniqueness of the individual and affirming their value inherently involves attending to their subjective experiences and a recognition that these experiences are particularly those of the subject to which an outsider can have no direct or objective access.

In this chapter, I have explained why respect is important even from an autonomous patient's perspective. The very experience of illness is itself objectifying and patients can experience a sense of alienation from their bodies. For patients the experience of being respected can make the disruptions to the lived body more bearable. The shift of attention involved in respect parallels the difference between healing and mere treatment or even cure. In treating or curing the body one takes an objective stance. In contrast, the interactive second-person perspective of respect helps to heal the rift in embodied agency. To be successful in healing, one needs to address not just technical remedies but the "global sense of disorder that permeates the patient's everyday life" (Toombs 1995, p. 20). Respect is important for patients because it can help restore a first-person sense of subjective agency. Since my account of respect involves treating the other as more than a mere object by affirming their equal moral value and attending to the subjective experiences of those in the health care encounter, one might wonder whether another concept such as sympathy or empathy might be more appropriate than the concept of respect. For example, I have drawn from Havi Carel's phenomenological discussion of illness, and she suggests that what is needed in health care is more empathy for the first-person experience of illness (2013a, p. 45; 2013b, p. 194). In this section, I explain why it is important to think about respect in particular as opposed to a different concept such as empathy or sympathy.

There are a number of accounts that elucidate the concepts of empathy and sympathy and philosophers disagree with one another about the best understanding of these concepts. In this section, I rely on the account of empathy and sympathy articulated by Stephen Darwall (1998, 2006) because he clearly connects these concepts to respect: a connection that I believe is mistaken. According to Darwall, empathy is an emotional experience from which I take up the position of another, whereas sympathy is an emotional experience in which I appreciate another's perspective from my own perspective (1998, 264). Sympathy, he writes, takes a third-person perspective on the situation and focuses on "the other and the relevance of her situation *for her*" (1998, 270; italics in original). In contrast, empathy is a way of occupying the other's perspective that involves imagining her situation from her perspective. Darwall says that when we empathize with another we assess the propriety of the other's feelings and when we find them appropriate to the situation, we "second" that feeling by feeling with the other. When we do not think the other's emotions are appropriate it is more difficult to empathize with them, which is itself an expression of our view that the other's feelings are unwarranted (1998, 269).

Elsewhere I have addressed some more general concerns with making empathy a part of a conception of respect (Schwartz 2011), but here I will focus on empathy in health care contexts in particular.

Empathy seems to involve feeling with others, or taking their emotions on board. But I am not convinced that respecting the unique particularity of the patient and their experiences requires *feeling with* the patient. In addition, given the often traumatic nature of medical encounters, it seems like asking health care professionals to feel with their patients might be harmful to the health professionals. Many of the emotional experiences of patients and their families involve difficult emotions, such as fear, loss, sadness, confusion. There is currently an under-appreciation of the emotional labour engaged in by health care professionals and I do not think it is wise to suggest more of this labour be undertaken as a moral *requirement* of the job. I do not think that it is bad if health care professionals empathize with their patients, and some feelings of empathy might be unavoidable. I fear, however, that if we recommend that health care professionals *ought to feel with* their patients this could lead to emotional problems for the health care professionals. As Richard Sennett writes, "a doctor… cannot let himself go to pieces whenever a patient dies; the doctor faces a truly overwhelming reality, and needs reserve as self-protection, in order to keep on operating" (2003, p. 146, loc. 1957). Further, it is not obvious that feeling with patients will be beneficial to their treatment or outcomes. For example, most health care professionals avoid treating their own family members because their empathy might cloud their judgement. If health care professionals were asked to care about and feel with all of their patients, there might be a similar distortion in medical decision-making.[6]

Although I have argued that respect will involve attending to the non-fungible subjective experiences of patients, this does not involve feeling with them, and in particular it is important not to think that this means personal intrusions. Richard Sennett (2003) argues that one of the central aspects of respect is that it recognizes the importance of impersonality in many of our public interactions. Sennett suggests

---

[6]Empathy might also lead to unequal treatment that would be troubling in health care settings. Baston et al. found that when study participants empathized with patients they were more likely to allow the person to jump the queue (cited in Prinz 2011, p. 228). Health care decisions ought to be based on fair procedures and policies rather than on the vagaries of our emotional configurations.

that there is a certain kind of compassion (or worse, pity) that can be found among public servants, which wounds the self-respect of those dependent on social services (2003, p. 133, loc. 1768; p. 136, loc. 1820). Citing Hannah Arendt, Sennett believes that if the purpose of social institutions is to do the recipient good, the feelings of the caregiver should be irrelevant (2003, p. 139, loc. 1856). Respect allows an impersonal caregiving that is neither indifferent nor unkind. Young's (1997) account of asymmetrical reciprocity that I describe last chapter involves careful attention to the expression of others across differences. Respect allows people with racial, ethnic, gender, class, ability and other differences to interact with one another on equal terms, but part of this equality involves recognizing their differences and opacity to us rather than empathetically collapsing those differences into a presumed sameness.

4.6 Conclusion

Although 'respect' is usually understood as a requirement to accept patient choice, I have argued that moral respect should additionally be understood as a requirement to attend to the subjective experiences of patients which can become objectified and alienated as patients experience the life changes associated with the phenomenology of illness. Moral respect as an attention to the subjectivity and embodied agency of the patient recognizes the inimitable uniqueness of each individual. Respect is important in health care contexts because scientific medicine involves an inherently objectifying aspect. The objectification intrinsic to scientific medicine might be part of its success; nevertheless, the ill-person must be able to recover their place as a subject, which can be difficult in the objectifying contexts of illness and medicine. I suggest that respect is important in health care ethics because it can counter the objectification of illness and medicine and being respected might help patients recover a sense of themselves as subjects rather than mere bodies undergoing hardship.

References

Baron, Richard J. 1985. An introduction to medical phenomenology: I can't hear you while I'm listening. *Annals of Internal Medicine* 103 (4): 606–611.

Bourdieu, Pierre. 1977. *Outline of a theory of practice*, trans. Richard Nice. Cambridge: Cambridge University Press.

Bourdieu, Pierre. 1990. *The logic of practice*, trans. Richard Nice. Stanford, CA: Stanford University Press.

Brodwin, Paul. 2013. *Everyday ethics: Voices from the front line of community psychiatry*. Berkeley: University of California Press.

Carel, Havi. 2013a. *Illness: The cry of the flesh*, rev. ed. Durham: Acumen Publishing.

Carel, Havi. 2013b. Bodily doubt. *Journal of Consciousness Studies* 20 (7–8): 178–197.

Carel, Havi. 2014. The philosophical role of illness. *Metaphilosophy* 45 (1): 20–40.

Darwall, Stephen. 1998. Empathy, sympathy, care. *Philosophical Studies* 89 (2–3): 261–282.

Darwall, Stephen. 2006. *The second-person standpoint*. Cambridge, MA: Harvard University Press.

Ells, Carolyn. 2001. Lessons about autonomy from the experience of disability. *Social Theory and Practice* 27 (4): 599–615.

Kontos, Pia. 2005. Embodied selfhood in Alzheimer's disease. *Dementia* 4 (4): 553–570.

Ley, Madelaine. 2017. Haptic conversations: Touch in medical care. Master's Major Research Paper. Ryerson University, Toronto, ON.

Merleau-Ponty, Maurice. 2012. *Phenomenology of perception*, trans. Donald Landes. New York: Routledge.

Prinz, Jesse. 2011. Against empathy. *The Southern Journal of Philosophy* 49 (s.1): 214–233.

Scheman, Naomi. 2009. Narrative, complexity, and context: Autonomy as an epistemic value. In *Naturalized bioethics: Toward responsible knowing and practice*, ed. Hilde Lindemann, Marian Verkerk, and Margaret Urban Walker, 106–124. Cambridge: Cambridge University Press.

Schwartz, Meredith Celene. 2011. Respect and health care ethics: Respect, social power, and health policy. Doctoral dissertation. Dalhousie University, Halifax, NS. https://dalspace.library.dal.ca/bitstream/handle/10222/14366/Schwartz,%20Meredith,%20PhD,%20PHIL,%20November%202011.pdf?sequence=3.

Scully, Jackie Leach. 2008. *Disability bioethics: Moral bodies, moral difference.* New York: Rowman & Littlefield.

Sennett, Richard. 2003. *Respect in a world of inequality.* New York: W. W. Norton (Kindle e-book version).

Toombs, S. Kay. 1988. Illness and the paradigm of lived body. *Theoretical Medicine and Bioethics* 9 (2): 201–226.

Toombs, S. Kay. 1995. The lived experience of disability. *Human Studies* 18 (1): 9–23.

Young, Iris Marion. 1997. Asymmetrical reciprocity: On moral respect, wonder, and enlarged thought. *Constellations* 3 (3): 340–363.

Young, Iris Marion. 2005. *On female body experience: "Throwing like a girl" and other essays.* New York: Oxford University Press.

5

Conclusion

Abstract Schwartz offers a summary of the egalitarian concept of second-person asymmetrical moral respect. Moral respect is grounded in dignity and affirms the equal moral value of each member of the moral community. The concept of moral respect is egalitarian because it is not grounded in contingent facts or capacities. Moral respect affirms a person's value as more than a *mere* object. Schwartz summarizes various forms of objectification described in the book. The account of moral respect has advantages because it is thoroughly egalitarian and can explain *why* the concept of moral respect is central in objectifying medical encounters. The discussion of moral respect is limited because the relations involved take time and material resources; they cannot be reduced to a checklist or set of tasks.

Keywords Moral respect · Moral dignity · Second-person standpoint · Objectification

© The Author(s) 2019
M. C. Schwartz, *Moral Respect, Objectification, and Health Care,*
https://doi.org/10.1007/978-3-030-02967-8_5

5.1   Introduction

My aim in this book has been to describe an egalitarian account of moral respect that applies to both non-autonomous and autonomous patients. In addition, this account should explain *why* moral respect is central in medical contexts. In order to meet these goals I shifted my focus from concerns about the target of respect (or *what* ought to be respected) to a consideration of the role of respect in moral life generally and in health care contexts in particular. I have argued that the particular role played by moral respect is the recognition and affirmation of the equal moral value of members of the moral community. When we respect an other we see them as irreplaceable. Moral respect is grounded in dignity, so the presence of this kind of value tells us why moral respect is owed or warranted. Moral dignity is an intrinsic, inalienable, absolute value that contrasts with the extrinsic value of mere price. Because dignity attaches to unique, non-fungible beings, dignity is both a universal egalitarian moral value that also points toward the concrete particularities of one's subjectivity. Respecting another involves interactive relations that treat others in ways that affirm their dignity. Disrespecting another reduces them to a mere object.

This description of respect can also explain why respect is particularly important in medical contexts. Modern medicine is inherently objectifying. Experiences of illness also threaten our subjectivity and continued existence. Moral respect is important in health care contexts because it can counteract the objectification inherent in medical contexts. Health care practices, medical research and medical knowledge all involve an inherent pull toward an objective perspective, and respect reminds us that we must always, at the same time, attend to the subjectivity, individuality, and dignity of those who are receiving medical care or who are participating in research. Respect reminds us to humanize patients.

In Chapter 2, I suggested some ways that medicine, medical research, and medical knowledge, all have inherently objectifying aspects. Medicine involves, at least in part, adopting an objective attitude to consider the physiological functioning of a patient's object-body. The bureaucratic nature of complex, modern health care systems also adopts an objective attitude toward patients following them as case numbers.

The pressures and time constraints of these institutions can create incentives that further encourage health professionals to adopt an objective attitude toward patients. These aspects, taken together, can be deeply objectifying where a patient might come to feel as though they are being *reduced to a mere* object. We can adopt more than one perspective on ourselves and others, however. Strawson says that the objective and interactive attitudes are opposed to one another, but they are not exclusive of one another (1974, p. 9). The reason we need an account of moral respect in health care contexts in particular is to counter this objectification experienced by patients. Engaging respectfully from the interactive stance involves attending to a patient's uniqueness, their non-fungibility, and their inherent equal moral value (or dignity). To respect another is to recognize the other as more than a mere object among objects. Respecting another affirms that they are the locus of subjectivity, perspectives, values, affects, and experiences that are inimitably theirs. Respect is important in health care contexts to ensure that the objectification of a patient that is inherent to health care does not become the primary or only way in which we interact with patients. Respectful relationships create the context in which medical objectification is tolerable for patients.

I described ten forms of objectification that can occur in health care contexts: instrumentality; denials or violations of autonomy; inertness; fungibility; violability; ownership; denials or violations of subjectivity; reduction to the body or diagnosis; silencing; and alienation. Respectful relationships can counter these forms of objectification and render them morally permissible because respecting another involves an interactive engagement from a second-person perspective that affirms a person as more than a mere object. The irreducibility of our individuality is expressed in our subjectivity our embodied agency, and our autonomy. Respect is the recognition of the inalienable, absolute, moral value of each unique individual. When we respect an other we see the other as non-fungible and irreplaceable. This is in contrast to one of the perspectives in medicine that considers the object-body as interchangeable with other bodies. Attending to the uniqueness of the individual and affirming their value inherently involves attending to their subjective experiences and a recognition that these experiences are particularly those of the subject to which I can have no direct or objective access.

Engaging respectfully with our patients involves adopting a different kind of perspective in our relationships than that suggested by health care practices that consider the body as an object of treatment or cure.

In Chapter 3, I argued that respecting non-autonomous patients is possible if we understand respect as grounded in dignity and broaden the account of second-person respect to include engagements beyond reactive attitudes. Reactive attitudes have only a limited role in health care contexts, even for patients with full autonomous capacities. In addition to reactive attitudes, second-person interactive relations can involve an asymmetrical care-respect. Care-respect occurs from the second-person perspective when it takes the cared-for themselves as the object of care, rather than taking an objectively determined criterion of welfare as the object of care. This kind of care requires attending to the response to care from those cared-for rather than simply relying on measures of improved welfare.

Care-respect is an asymmetrical form of reciprocity for two reasons. First, we cannot assume a symmetry of perspectives so that I can merely empathetically project myself into your position to decide whether the care-respect has been well received. The differences among even autonomous persons are too great for imaginative projection to be an effective means for judging responses to care. The differences in perspective are even more significant when a person's subjective experiences are affected by illnesses which we have not experienced from a first-person perspective. Instead, we should attend to the response of the other. Sometimes this will take the form of verbal exchanges between two autonomous persons who can take the words at their surface meaning. In other cases, where the illness in question involves significant cognitive impairment or changes in one's subjective experience then the attention required might have to go beyond the surface meaning of the words to find their deeper meaning. For those who are non-verbal, attention to body language, emotional expressions, or embodied agency will be more important.

Second, care-respect is asymmetrical because there is a difference between what is given and what is returned. Symmetrical accounts of respect assume an equal authority so what is given by one can be returned by the other. This form of symmetrical reciprocity resembles market exchanges where obligations of respect are exchanged without remainder. In contrast, asymmetrical reciprocity is more akin to gift-giving in which what is given differs from what is returned. Similarly, in care-respect the reciprocity of respect will take different forms that vary with the capacities of those engaged in the relation. An infant may not be able to offer explicit reasons, but even an infant can engage in the mutuality of relations and respond to the care received.

In the fourth chapter I argued that this account of asymmetrical care-respect that counters the inherently objectifying elements of modern medicine would be important even for autonomous patients experiencing serious illnesses. The phenomenological experience of illness often includes a shift of attention to the object-body. Even autonomous patients would benefit from second-person interactions that affirm their subjectivity. This can help patients live well with their illness rather than feeling alienated from their own body.

In this conclusion I tie up some lose ends, forestall a potential objection, and consider the advantages and limitations of this account. In particular, I return to the assertion that respect can 'balance' the objectification involved in medicine and the experience of illness. In this conclusion I describe some criteria for identifying when balancing objectification will be more significant. In Sect. 5.3, I describe some of the advantages of the account. Then, in Sect. 5.4, I turn to a potential objection to the overall project. One might object that the concept of 'autonomy' could be expanded to cover the concerns I have discussed. I provide the reasons I do not think it would be preferable to expand the account of autonomy. In Sect. 5.5, I conclude with some of the limitations of my account. In particular I consider time constraints and constraints of institutional organization. I reconsider the cases I introduced in Chapter 1 and I describe how my account would make a difference in these cases.

5.2   Moral Respect and Balancing Objectification

In Chapter 2, I described respectful second-person relations as a balance among perspectives. I have suggested that the objectification involved in medicine is not inherently bad and has led to many important medical successes. However, objectification can become the dominant mode of engagement and this can lead patients to feel dehumanized. I made the analogy to CO_2 in the atmosphere, which is not inherently bad and is even important for supporting plant life, yet can nevertheless get out of balance. In Chapter 3, I described some ways that objectification can be balanced by asymmetrical care-respect. Chapter 4 considered some phenomenological accounts of when this balance will be important for people experiencing illness.

In this section, I pull these threads together to suggest some criteria we could use to decide when the balancing will become important. In some medical encounters the forms of objectification will be minor or temporary and so adequately balanced by standard procedures like informed consent. So when would a more robust second-person care-respect be called for? This form of respect would be particularly important when many forms of objectification are present, when the illness is life-changing, or when the person's resources for countering their own objectification are limited either as a result of a lack of autonomous capacities or as a result of institutionalization which limits the availability of traditional supports.

The first, and most straightforward, case would be one in which the patient lacks the capacity to make autonomous decisions. In such cases only moral respect that affirms their dignity can directly respect the patient as they are unable to provide informed consent. A second criterion for deciding when balance will be important has to do with whether the illness can be expected to be experienced as life-changing. A minor illness like the flu, or even something more serious but temporary like an appendectomy, many not require additional attention. In contrast, serious illnesses, like cancer, that are potentially fatal involve greater existential threats. Conditions that alter a person's life-plans, like infertility or hysterectomy for those who wanted to be parents, will also involve greater disturbances to one's bodily confidence. Which conditions will alter life-plans will depend on the particular patient. For example, some people might be infertile but never wanted children and so the experience might involve little existential change. The second criterion will be patient-relative. When the illness does involve changes to one's life plans or existential threats then these conditions can be expected to involve more extensive forms of objectification and alienation.

Third, chronic conditions that involve ongoing debilitation or deterioration provide the time for an ongoing respectful engagement. Unlike emergency situations that call for an immediate response, where that response is appropriately focused on the object-body, chronic conditions allow for engagement over time. Finally, if a patient is institutionalized, then, a balancing respectful engagement will be particularly important. When we are in an institutional setting we lose access to those things that typically personalize us. Even though our relationships to other people may continue, the institution will set limits on when friends or family can visit. While we are in an institution we lose contract with the objects in our homes that usually surround us and remind us who we are. In our homes the placement of objects reflects our patterns of living. In contrast, in an institutional setting objects are placed where it is most convenient for staff (Young 2005, pp. 156–157). Even temporary institutionalization will involve disruptions or losses of contract with the people and objects that typically support our sense of self. In institutional settings it will be more important for health care professionals to help support a patient's sense of self to counter this loss of external supports.

These four criteria can combine and the more there are present the more important it will be to help a patient retain a sense of their subjectivity and dignity. The idea of 'balancing' the objective and interactive stances also allow us to consider whether there could be a similar negative imbalance on the side of too much second-person engagement. Perhaps too much second-person interaction or too much attention to a patient's subjectivity or uniqueness would also result in morally troubling situations. There is reason to believe such problems might arise. Strawson suggests that adopting a wholly objective attitude toward someone precludes adopting reactive attitudes toward that person. Reactive attitudes include such things as resentment, anger, and so forth (1974, p. 9). If there were an imbalance toward too much of the interactive attitude, this could create a situation in which a health care professional might be inclined to allow reactive attitudes to be inappropriately expressed toward a patient. Professional distance and objective attitudes likely have an important role to play in preventing inappropriate interactions.

For health care professionals there might be advantages of the objective stance. Strawson notes that objective attitudes can be a source of refuge, when the strains of involvement are too great. Given the often emotionally charged nature of medical care and the attendant worry, fear, and vulnerability of patients and their families, it might sometimes be valuable to adopt an objective perspective. But Strawson also suggests a balance is needed when he writes, "Being human, we cannot, in the normal case, do this for long, or altogether. If the strains of involvement, say, continue to be too great, then we have to do something else—like severing a relationship" (1974, p. 10). Health care professionals are asked to provide too much second-person interaction if they are required to empathize with each patient's emotionally draining experience. But it also might be inhuman to expect health care professionals never attend to their affective experiences. Requiring only an objective robotic interaction could equally lead to caregiver burn-out.

5.3   Advantages of the Account

There are some advantages to focusing on the role of respect as balancing objectification in health care. Although the concept of respect is frequently invoked by bioethicists, we rarely articulate *why* respect is important in medical contexts in particular. Instead, its importance is more often assumed or simply asserted. Bioethicists sometimes suggest that respect is important in bioethics to protect patients against the risk that health care professionals will exploit patients for their own ends or encourage the patient's dependency, but they do not explain *why* this temptation to foster dependency might arise in health care. There is no explanation for why *researchers, physicians,* or other *health care professionals* might be in danger of encouraging dependence among their patients. The temptation to exploit patient vulnerability might come from the existence of conflicts of interest. Although health professionals often have the interests of patients in mind, they are also working with competing interests, such as their own academic advancement or the financial interests they might have as the result of various incentives or investments in treatments under study.

I provide a slightly different view of why respect is important in health care. The account of respect that I have offered explains why respect will be important in medical contexts in particular without impugning the character of health care professionals. Respect is not primarily an important principle because it counters a health care professional's exploitation of vulnerable patients or their tendency to render them dependent. Instead, respect is important because it counters the objectifying elements of medicine and illness. The account of respect that I offered explains the importance of respect in health care as a feature, not of the health care professionals, but of health care (and scientific) practices in complex, bureaucratic institutions. Respect is important regardless of the character of the individual health care professional because medicine, research, and the experience of illness and care in modern institutions are all objectifying, both in the attempt to be unbiased and, as a consequence, by treating the ill body as a fungible physiological object. Respect is important because it reminds us to attend to the subjective, non-fungible, particular experiences of the patient.

I have argued that the role of respect in bioethics is to counter the objectification inherent in health care contexts. Creating respectful relationships involves attention to the subjectivity of a patient that recognizes the inimitable uniqueness of each individual. Respect is important in health care contexts because scientific medicine involves an inherently objectifying aspect. The objectification intrinsic to scientific medicine might be part of its success; nevertheless, the ill-person must be able to recover their place as a subject, which can be difficult in the objectifying contexts of illness and medicine. I suggest that respect is important in bioethics because it can counter the objectification of illness and medicine and being respected might help patients recover a sense of themselves as subjects rather than mere bodies undergoing hardship.

One way of being respectful toward those who are competent is to acknowledge their decision-making and their right to determine the course of their own lives; that is, respecting autonomy is often part of being respectful toward patients. So, the argument I am providing does not imply that respect for individual autonomous decision-making is unimportant. For those who are competent, that understanding of respect will remain an important component of respecting *them*. But if respect is to play a robust role in countering the various forms objectification might take in medical contexts respectful relationships will involve more than a narrow focus on respecting the decisions of autonomous patients.

A second advantage of the account of respect in this book is that it can include non-autonomous patients on the same terms as autonomous patients. Beauchamp and Childress stress that when they use the term 'autonomy' they are referring to the particular *choice* being made and not to broader features of the person or their ability to self-govern (2009, p. 100; 2013, p. 102). Because of this focus on choice, mainstream bioethics has often used the criterion of 'competence' in a gatekeeper function to separate those whose autonomous choices must be respected from those whose choices are not made intentionally or with sufficient understanding and freedom from controlling influences. While one might have to be competent to make a sufficiently autonomous choice, even those who lack the capacity for autonomous choice (and possibly those who are in a coma) will nevertheless have subjective experiences, they will remain unique and non-fungible individuals, and they deserve to be treated as more than a mere object among objects. By focusing on the role of respect as countering the objectification inherent in health care we can think more deeply about how to engage respectfully with non-autonomous patients, as described in Chapter 3.

While respect for autonomy only applies to a subset of patients (those who are competent) all patients including those who lack the capacity for choice will have subjective experiences, they will be unique and non-fungible, they will be bearers of equal moral value. In the seventh edition of *The Principles of Biomedical Ethics*, Beauchamp and Childress make clear that the principle of respect for autonomy does not extend to those who are not sufficiently autonomous and cannot be rendered autonomous. Those who are not sufficiently autonomous still have moral standing, they write, and so are still owed moral respect (2013, p. 108). They do not, however, expand on what this moral respect might look like. I suggest that moral respect in medical settings will involve creating respectful relationships that help to balance the objectification present in health care. I have argued that the moral work that is done by a concept of respect is to prevent medical epistemology and bureaucratic medical systems from considering a patient merely as an object to be treated, managed, or cured. One of the central functions of respect is to recognize something or someone as more than a mere object among objects. The concept of respect is important in health care ethics for that reason.

5.4   Moral Respect, Objectification, and Autonomy

It might be objected that respect for autonomy would cover the concerns I mention above, perhaps especially when it is combined with the principles of beneficence, non-maleficence, and justice. Some might think that many of my concerns could be met by an expanded account of autonomy, perhaps drawing on accounts of relational autonomy, or alternatively incorporating phenomenological accounts of lived autonomy into the discussion. So why would I suggest we focus on the role of moral respect in countering the various forms of objectification? Perhaps a more robust process of informed consent could address the issues I have raised.

Informed consent and respecting a patient's autonomous decisions are important *aspects* of engaging respectfully with competent patients. But informed consent alone is not sufficient for engaging respectfully with even competent patients in the robust way that is required for countering the various forms of objectification in health care. Informed consent discussions can sometimes become rigid and standardized as lawyers advise about what needs to be disclosed in these discussions. The informed consent "discussion" might be reduced to an information sheet handed to a patient with an offer to answer any questions that might arise, as in my experience described as *Case C* in Chapter 1. When informed consent becomes another task for the health professional to complete, the patient might experience the process as a further objectification rather than a respectful engagement that counters it: the information that is disclosed during the informed consent process focuses on the risks and benefits that are garnered from generalized medical research; the patient is then asked to consider themselves as bodily-objects in relation to the other bodily-objects on whom these studies were performed. Of course, the patient is also considered as a subject, an agent or as autonomous when they are allowed to evaluate the risks and benefits associated with different treatments in relation to their own values and preferences in order to make a decision about treatment. But treatment decisions are again about interacting with a patient's body considered mainly as an object of treatment.

In addition, I have concerns about expanding the account of 'autonomy' at use in bioethics. The principle of 'respect for autonomy' is well-entrenched in many legal systems within health law. This creates a clear-cut set of requirements that I believe ought to be maintained and strengthened. The legally entrenched form of autonomy tends to focus specifically on informed consent and respect for patient choices. I worry that expanding the concept of 'autonomy' in order to try to capture all of the various forms of objectification in medicine might weaken legal requirements or render them less clear. The kinds of failures of respect that I have been describing are morally significant but most of them are not legally actionable. I have included violations or denials of autonomy in the account, so some forms of objectification are legally actionable, but others are not. For example, the daughter with Down syndrome in *Case A* was reduced to a diagnosis and her presence in the room was ignored, but there was no legal transgression. Olive, from *Case D*, suffered from boredom as a result of her institutionalized care for Alzheimer's disease, but she was not abused. It is important to keep a clear account of 'respect for autonomy' that reflects its legal entrenchments. The concerns I have raised are moral wrongs but many of them do not rise to the level of legally actionable wrongs. My account remains connected to violations or denials of autonomy and informed consent because for many patients these are particularly salient forms of objectification.

On my account, informed consent remains important because for many patients, the experience of having one's will neglected or overridden is excruciatingly objectifying and completely at odds with being respected. Neglect of one's agency and autonomy contributes to the sense that one is being considered *merely* as an object among objects. Failing to get informed consent *is* sufficient for objectifying the competent patient and ignoring their subjectivity, uniqueness and inviolability. One need not be competent to deserve respect, although competence is a relevant criterion when deciding whether following a patient's decision is likely to be safe. If the patient lacks capacity, then the decision might be based on incomplete understandings or irrelevant considerations. Competence should not be thought of as dividing those whose decisions need to be respected from those whose decisions can be disrespected. The language of "respecting" patient choices seems to me to be misleading. A more precise way of discussing these issues would be to say we are "affirming," "accepting" or "following" the patient's decision. When bioethicists have discussed the 'gatekeeping' role of competence, they say that the decisions of competent patients must be respected, but they rarely suggest that the choices of those who are not autonomous should be disrespected. Both autonomous and non-autonomous patients should be respected and their decisions should be acknowledged. However, only autonomous patients have the capacity to make decisions that can be legally binding.

In this section I have shown why the role moral respect plays in medical contexts is not reducible to respect for autonomy, although respect for autonomy will remain one important element of creating respectful relationships with one's patients and the family members of non-autonomous patients. Although respecting a patient's autonomy, providing information to get a legitimate, free consent remains important on my account, it is not sufficient for creating a fully respectful relationship.

5.5   Limitations

One significant limitation of the account that I have described in this book is that it is not a form of respect that can be easily adopted by a health care professional in the absence of institutional supports. Tronto describes care as involving four phases: caring about, taking care of, caregiving, and care receiving (1993, pp. 106–107). Ideally these phases would be integrated so that responses to care attended to by caregivers would affect how we identify care needs and decide how to respond to them. In bureaucratic institutions, however, these responsibilities are often divided so those in managerial positions decide what to care about and how to meet those needs with little input from those providing care or receiving care. The policies put in place by those with little day-to-day caregiving experience can limit the ability of people charged with frontline caregiving to respond to those receiving care. The separations and divisions limit the effect that care responses have on decisions about what to care for and how to meet the care need. There needs to be institutional mechanisms in place for caregivers and care receivers to meaningfully engage with decision-making structures. These structures are not in place in many institutions. The account I have described does have some potential for effecting change, however. Most institutions affirm the importance they place on respecting patients. If their commitment is genuine, then implementing structural mechanisms to integrate the four phases of care has a moral justification.

A second limitation of this account of respect is that it might require more time than even a robust informed consent discussion and time is at a premium in most health care institutions. While it is true that the second-person interactions of respect take time they might also save time in some cases. For example, in *Case A* involving a mother seeking an ultrasound with her daughter, the informed consent discussion took a lot of time and created considerable frustration for both the mother and doctor. Simply asking the mother about her reasons for the test would have involved fewer repetitions of the same information and less miscommunication. Similarly, in *Case B*, the discussion of the importance of using birth control while undergoing chemotherapy could have been avoided entirely had the doctor simply asked Tara why she was not using birth control as had been recommended to her. This would have quickly revealed that the hysterectomy made birth control irrelevant and would have spared Tara the experience of being lectured to avoid motherhood, which was a lost possibility for her that she was in the midst of grieving.

In my *Case C* there was no lost time in an informed consent discussion. There was lost time because I went through the initial fertility work-up and then decided not to complete treatment. The initial consultations and tests led to no benefits and so time was lost from that perspective. Olive required residential care in *Case D*, so the time involved is somewhat consistent. She would still need to be fed, cleaned, and dressed. Allowing Olive to do these things for herself to retain her agency might take more time. Interacting with Olive rather than leaving her to watch television or look out a window would take time. But there might be time savings, too. For example, if greater respectful interactions leads to lower levels of frustration and fewer outbursts then this could save time dealing with the repercussions. It is not clear whether respectful engagements would ultimately take more time or less time; although this also seems beside the point. The issue is not simply about measures of time spent per patient. Perhaps more important are questions of whether patients and healthcare professionals find the respectful interactions more satisfying, humanizing, and fulfilling. The time required to create respectful relationships ought to be taken into account by those who propose that respect matters in their institutions. This means that those who make macro level decisions about health care systems, policy makers, and managers must consider the time and work involved in creating respectful relationships between health care providers and their patients when they make decisions about how the health care system, hospital or clinic operates. The work of creating respectful relationships requires material supports.

References

Beauchamp, Tom, and James Childress. 2009. *Principles of biomedical ethics*, 6th ed. New York: Oxford University Press.

Beauchamp, Tom, and James Childress. 2013. *Principles of biomedical ethics*, 7th ed. New York: Oxford University Press.

Strawson, Peter. 1974. *Freedom and resentment, and other essays*. London: Methuen.

Tronto, Joan. 1993. *Moral boundaries: A political argument for an ethic of care*. New York: Routledge.

Young, Iris Marion. 2005. *On female body experience: "Throwing like a girl" and other essays*. New York: Oxford University Press.

Index